Bio-Art to acc

THE WORLD OF BIOLOGY

Fifth Edition

Solomon & Berg

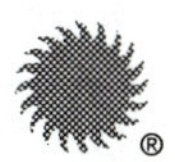

SAUNDERS COLLEGE PUBLISHING

Harcourt Brace College Publishers

Fort Worth Philadelphia San Diego New York
Orlando Austin San Antonio Toronto
Montreal London Sydney Tokyo

Printed in the United States of America.

Portions of this work were published in privious editions.

Saunders College Publishing; Bio-Art to accompany The World of Biology, 5E Solomon & Berg.

ISBN 0-03-05957-7

567 017 987654321

A NOTE TO THE INSTRUCTOR

Thank you for adopting **The World of Biology, Fifth Edition**. We appreciate your use of our text and as a special aid for students we have produced BioArt™.

Bio-Art™ is a collection of important pieces of art from **The World of Biology, Fifth Edition** rendered in black and white. Generally most pieces of Bio-Art™ do not include labels, so students can use the art as a learning tool. It is an excellent tool in measuring students understanding of processes and organisms.

This valuable study aid is free with copyright privileges to instructors who adopt **The World of Biology, Fifth Edition** for class or can be purchased by students at a low cost. Please contact your bookstore if you would like Bio-Art™ to be purchased by students as a required or recommended supplement.

Suggested uses of Bio-Art™ for instructors who adopt **The World of Biology, Fifth Edition**, include:

*Copying Bio-Art™ as handouts for students, enabling students to label parts of figures, take notes, and avoid redrawing complicated diagrams, while the instructor uses a corresponding color overhead transparency from **The World of Biology, Fifth Edition**.

*Duplicating Bio-Art™ on overhead acetates and referring to and writing on the acetates during lectures, while students refer to their own handouts.

*Distributing copies of Bio-Art™ as part of exams and quizzes.

*Having students complete homework assignments using their own copies of Bio-Art™

WORLD OF BIOLOGY 5E
by Solomon and Berg

BioArt list

1	4-2	Making and breaking polymers
2	5-12	Function of Golgi complex
3	5-19	Structure of a cilium
4	6-3	Structure of plasma membrane
5	8-3	Overview of aerobic respiration
6	9-6	Photosynthesis from different perspectives
7	10-5	Mitosis
8	10-7	Meiosis
9	13-6b	Structure of DNA
10	13-7	DNA replication
11	14-3	Structure of RNA
12	14-9	Elongation of a polypeptide chain
13	16-6	Identifying bacteria that have taken up genetically-altered plasmids
14	18-2	Founder effect
15	18-3	Bottleneck effect
16	18-4	Stabilizing, directional, and disruptive selection
17	19-11	Endosymbiont theory (origin of eukaryotes)
18	20-2	The five-kingdom system of classification
19	20-8	Cladogram
20	21-2	Events in a lytic infection
21	21-5	Structure of a typical bacterium
22	21-15a	Structure of *Euglena*
23	21-22a,b	Structure of *Paramecium*
24	22-2	Life cycle of black bread mold (zygomycete)
25	22-5a	Structure of basidiocarp and basidia
26	Table 23-2	Comparison of monocots and dicots
27	23-3	Moss life cycle
28	23-5	Fern life cycle
39	23-8	Gymnosperm life cycle
30	23-10	Flowering plant life cycle
31	24-3	Three basic animal body plans
32	24-4	Proposed evolutionary relationships illustrated by phylogenetic tree
33	24-20	Internal anatomy of a clam
34	24-24	Internal structure of an earthworm
35	24-30	Insect (grasshopper) body structure
36	25-7	Internal anatomy of the perch, a bony fish
37	26-1	Herbaceous plant body
38	26-3	External structure of a woody twig
39	26-14	Structure of a root tip
40	26-16	Wood and bark (secondary growth)
41	27-1	Cross section of typical leaf blade
42	28-7	Endodermis
43	28-12	Tension-cohesion mechanism
44	28-13	Pressure-Flow hypothesis

45	29-7	Overview of sexual reproduction in flowering plants
46	29-14	Embryonic development in shepherd's purse
47	30-2	Short-day and long-day plants' response to different photoperiods
48	30-3	Seed germination and growth of soybean (dicot)
49	30-4	Seed germination and growth of corn (monocot)
50	31-6	Principal organ systems
51	31-6	Principal organ systems
52	32-2	The structure of mammalian skin
53	32-6	The human skeleton
54	32-8	Skeletal muscle structure
55	32-10	Muscle contraction
56	32-14	Superficial muscles of human body, anterior view
57	32-15	Superficial muscles of human body, posterior view
58	33-1	Flow of information through the nervous system
59	33-5	Resting neuron & proteins in the plasma membrane form ion-specific channels
60	33-9	Transmission of an impulse along an axon
61	33-11	Transmission of an impulse between neurons
62	34-6	Comparison of brains of six vertebrate classes
63	34-7	The spinal cord
64	34-8	Lateral view of human brain
65	34-9	A midsagittal section through human brain
66	35-15	Structure of the human eye
67	36-6	Types of blood vessels
68	36-9	Section through heart
69	36-15	Circulation through some of the principal arteries and veins
70	36-16	The lymphatic system
71	Focus Figure	The evolution of the vertebrate heart
72	37-7	Antigen, antibody, and antigen-antibody complex
73	37-9	Cell-mediated immunity
74	38-6	The human respiratory system
75	38-7	Structure of the alveolus
76	39-4	The human digestive system
77	39-5	The wall of the digestive tract
78	39-7	Structure of the stomach
79	39-8	The wall of the small intestine
80	40-2	Osmoregulation
81	40-5	The human urinary system
82	40-6	Structure of the kidney
83	41-4	Principal endocrine glands
84	41-5	The hypothalamus secretes releasing & release-inhibiting hormones
85	41-6	Relationship between the hypothalamus and the posterior lobe
86	42-3	Male reproductive system
87	42-10	Anterior view of female reproductive system
88	43-11	Early human development
89	47-6	Hydrologic cycle

 Figure No. 4-2

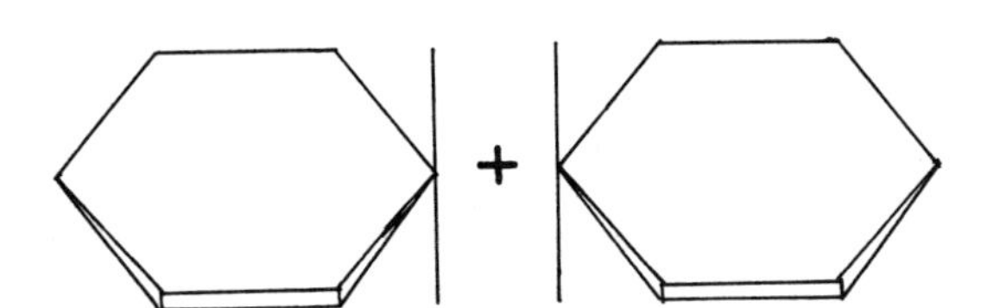

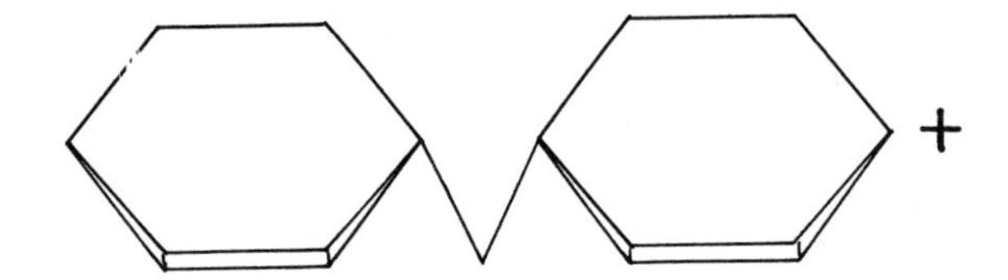

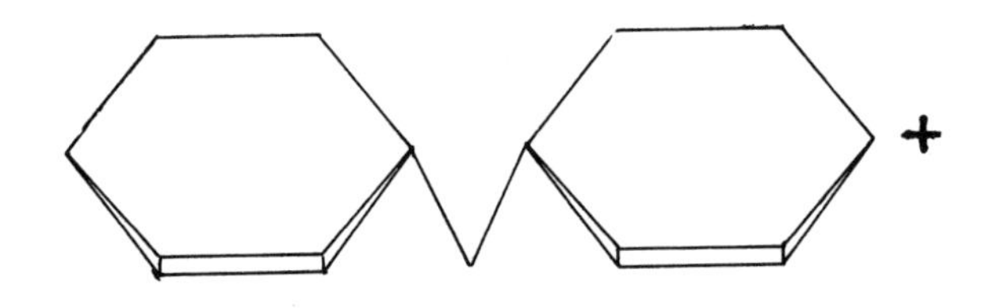

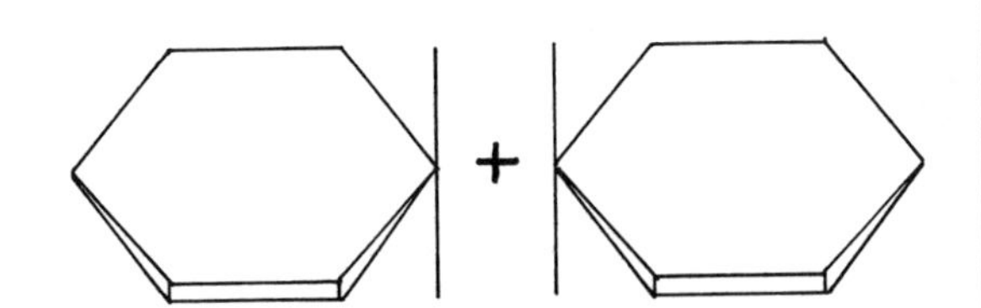

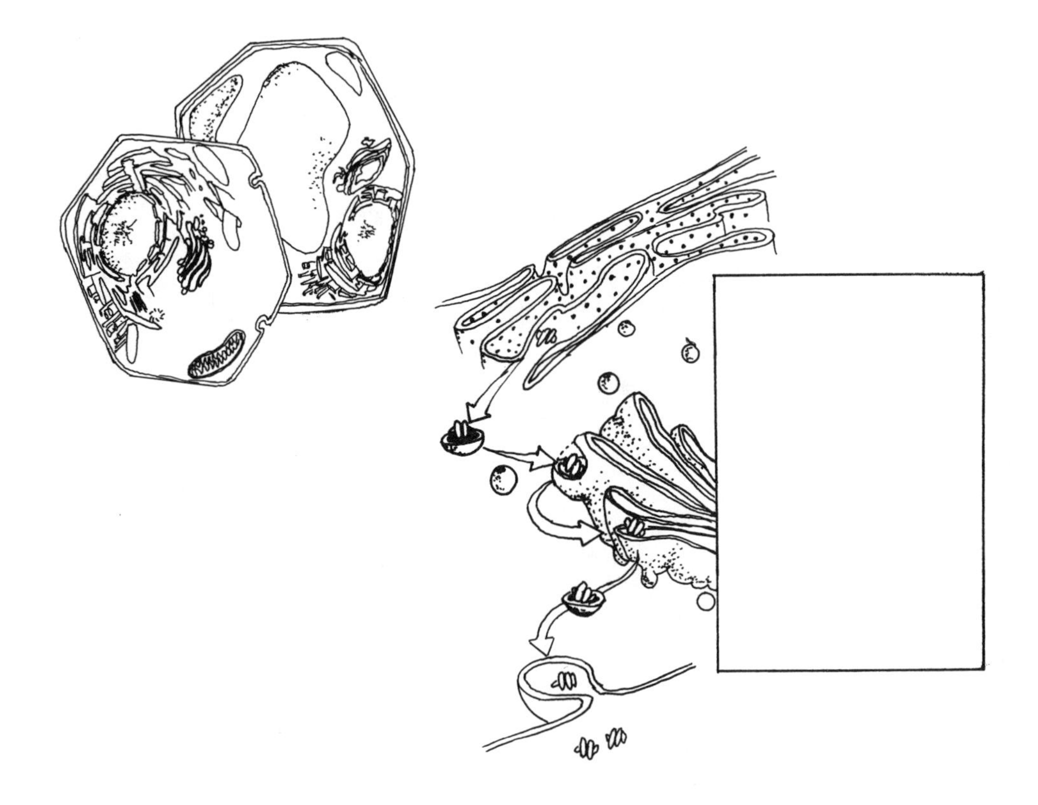

Figure No. 5-19

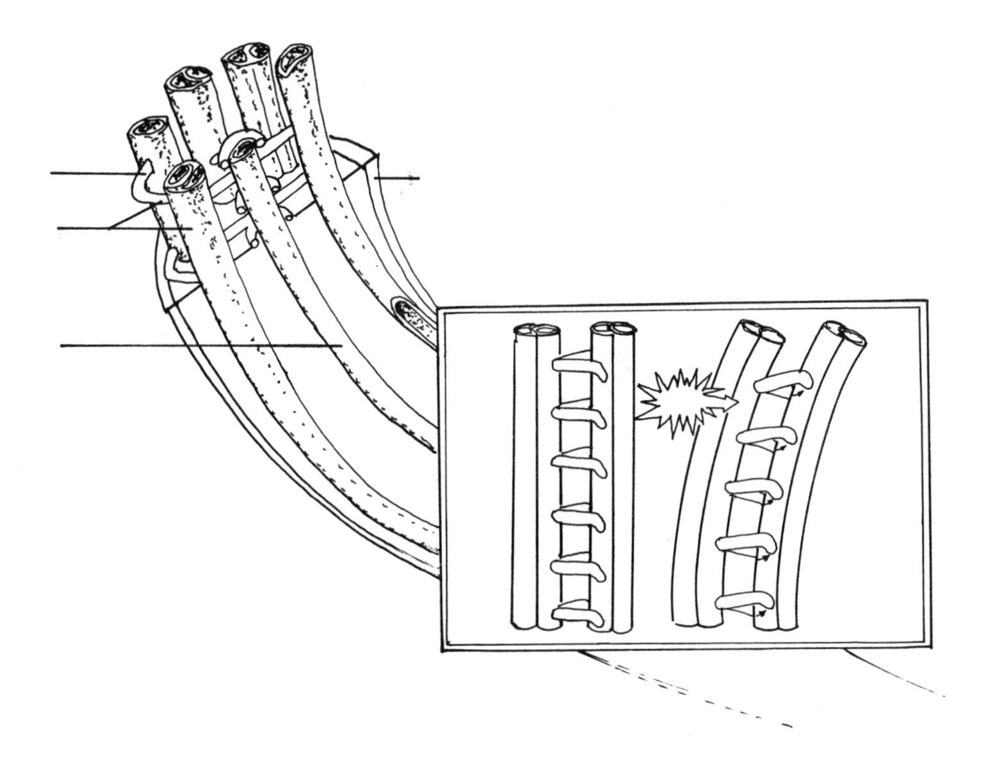

 Figure No. 6-3

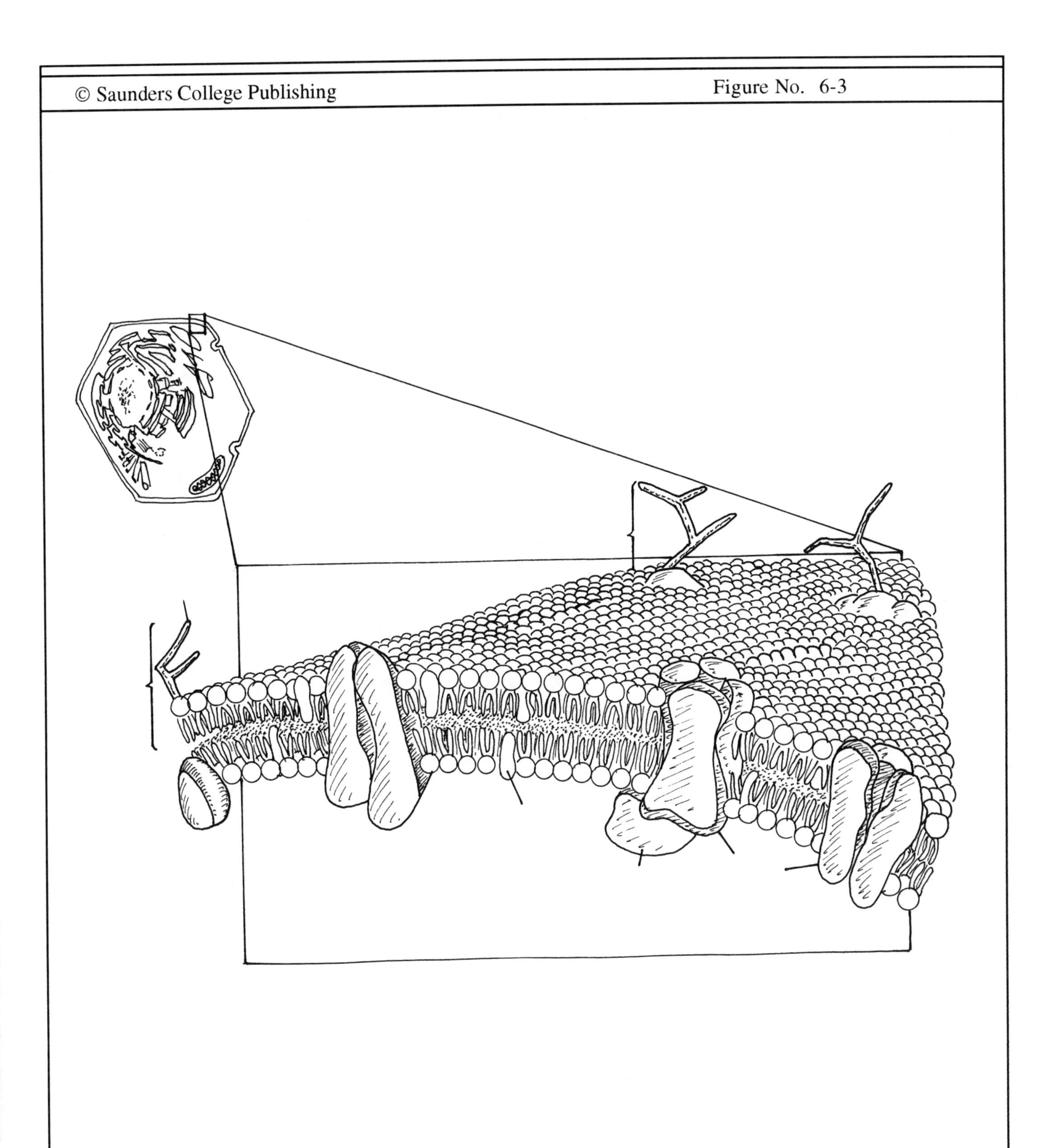

 Figure No. 8-3

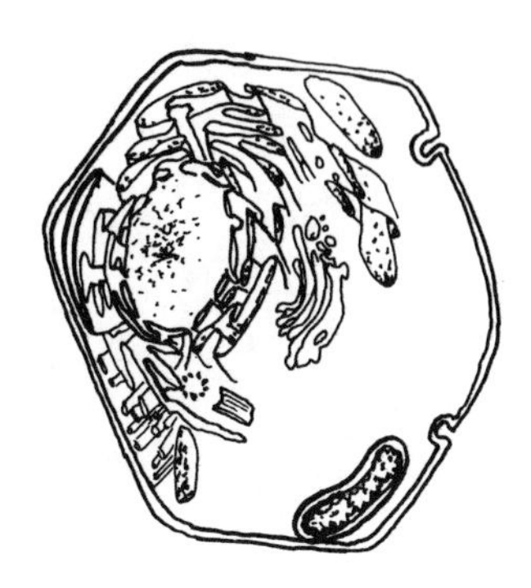

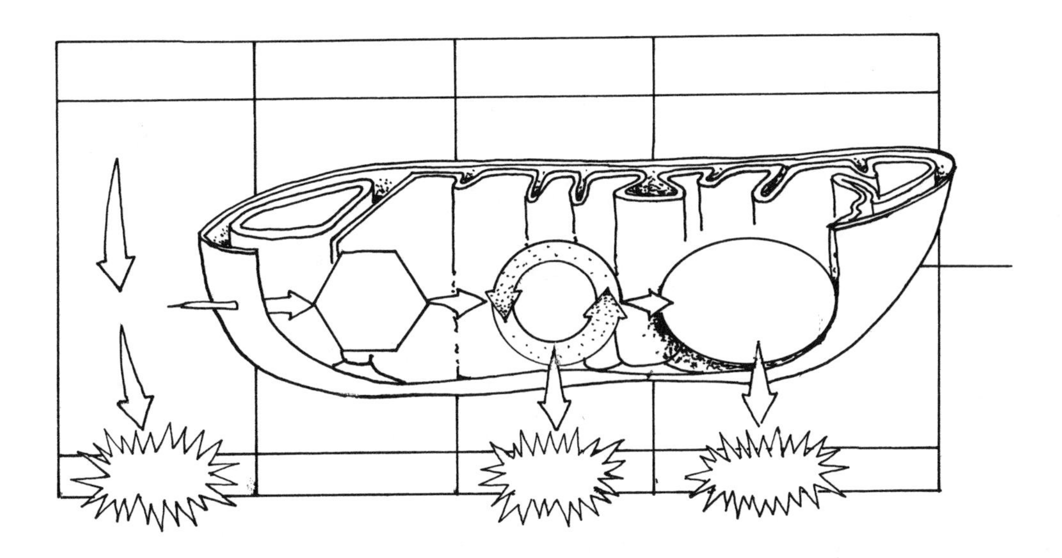

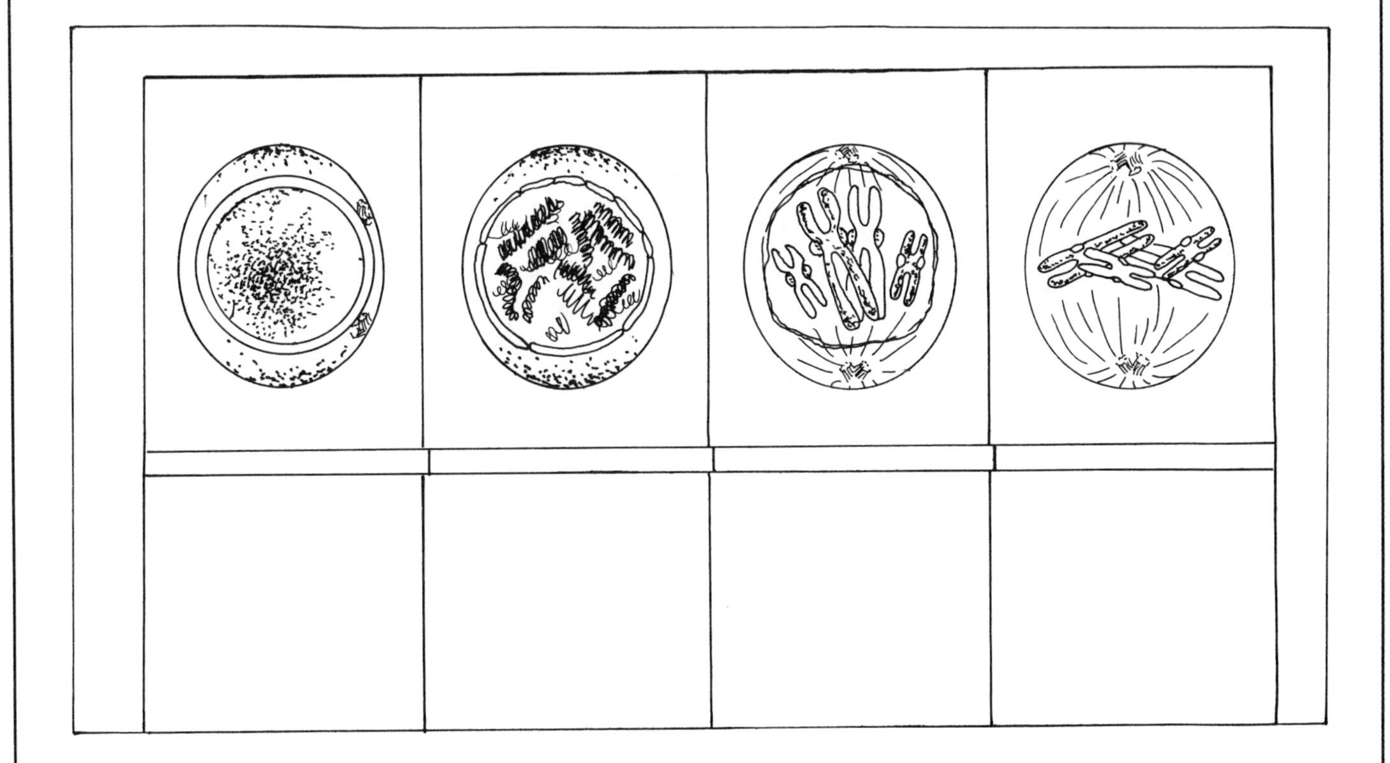

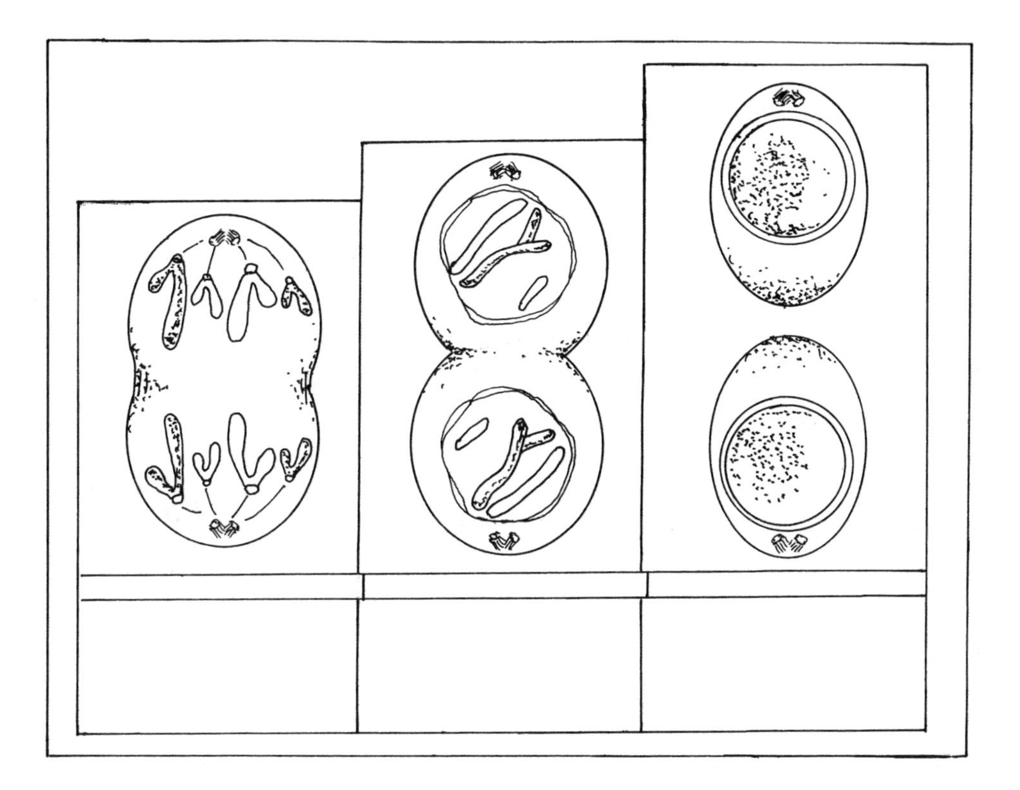

 Figure No. 10-7 pt 1

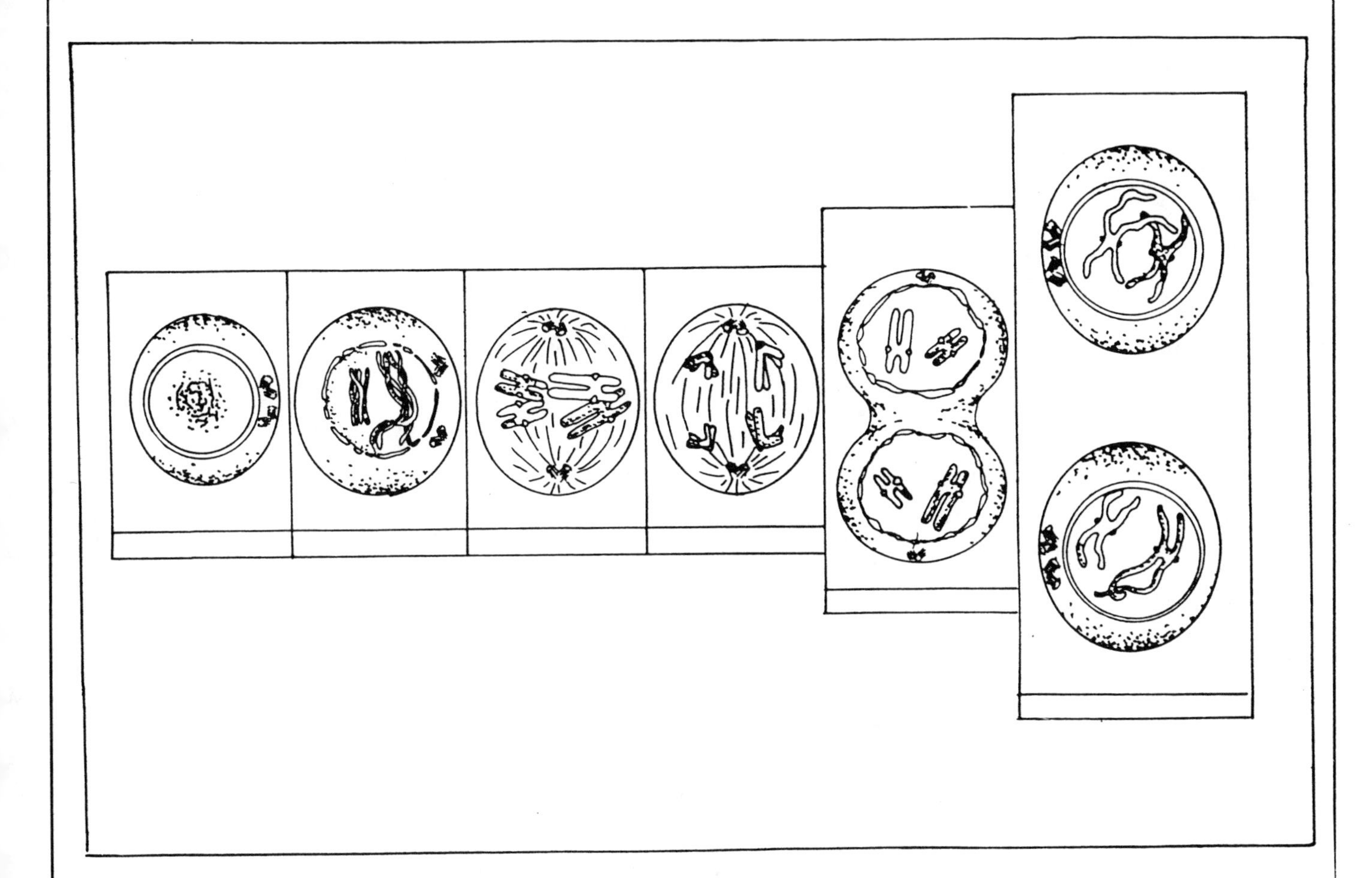

Figure No. 10-7 pt 2

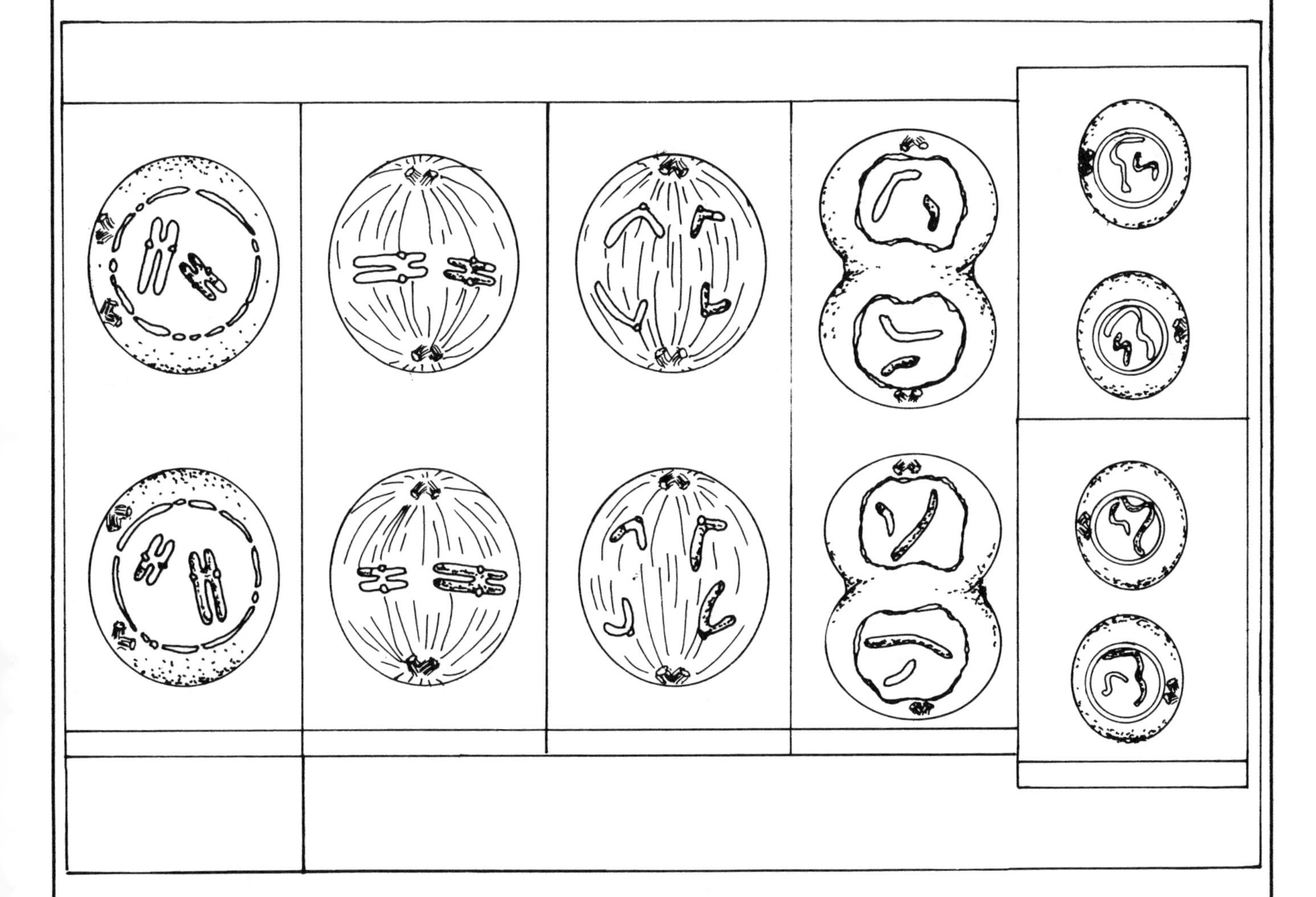

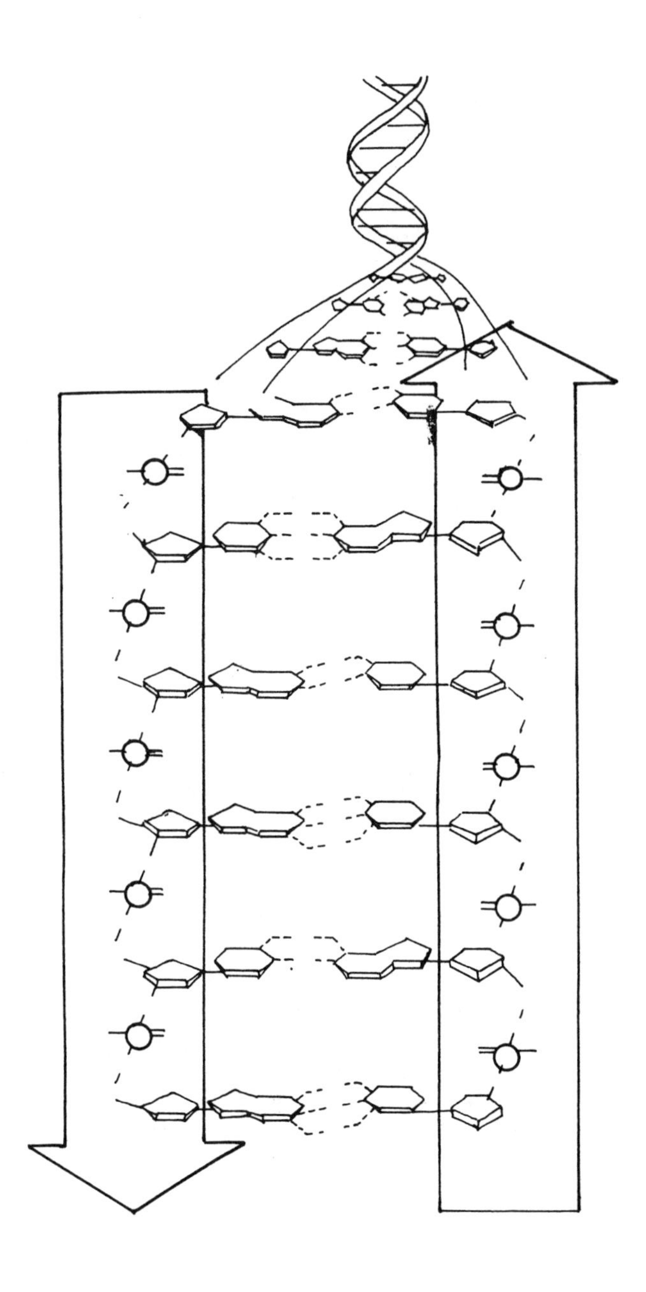

 Figure No. 13-7

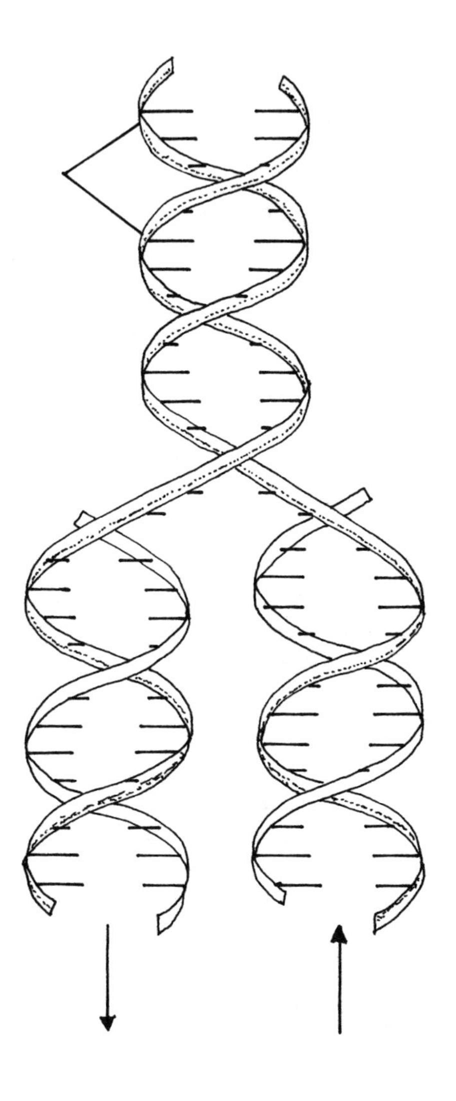

 Figure No. 14-3

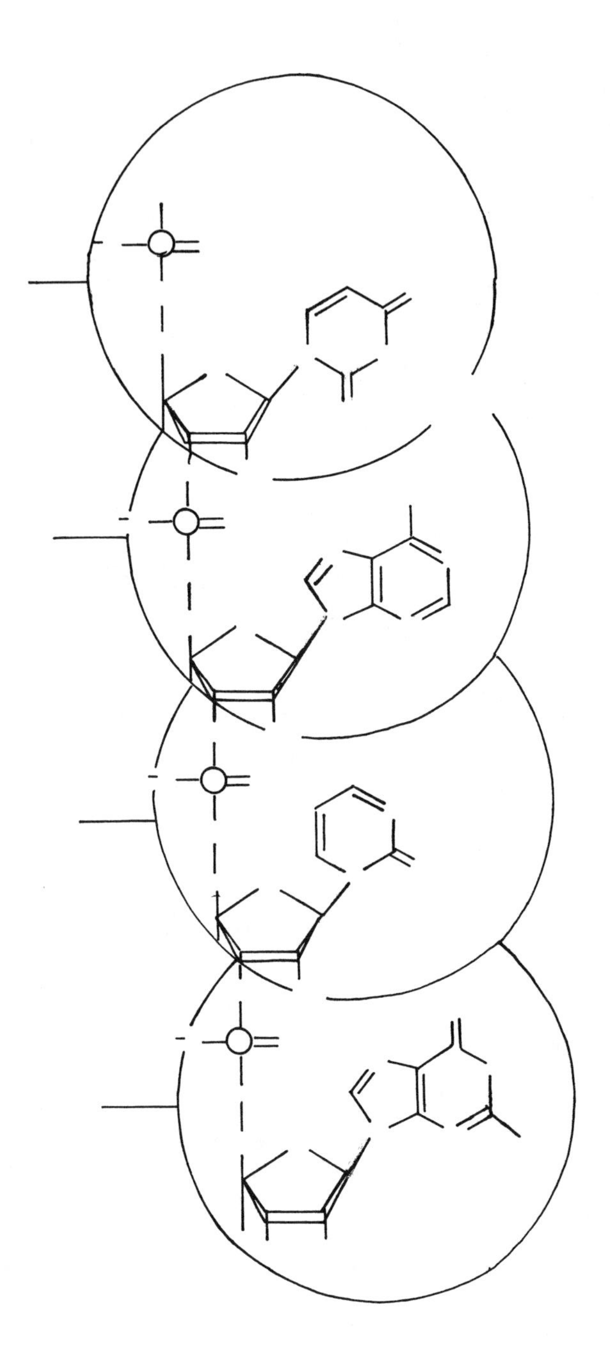

Figure No. 14-9

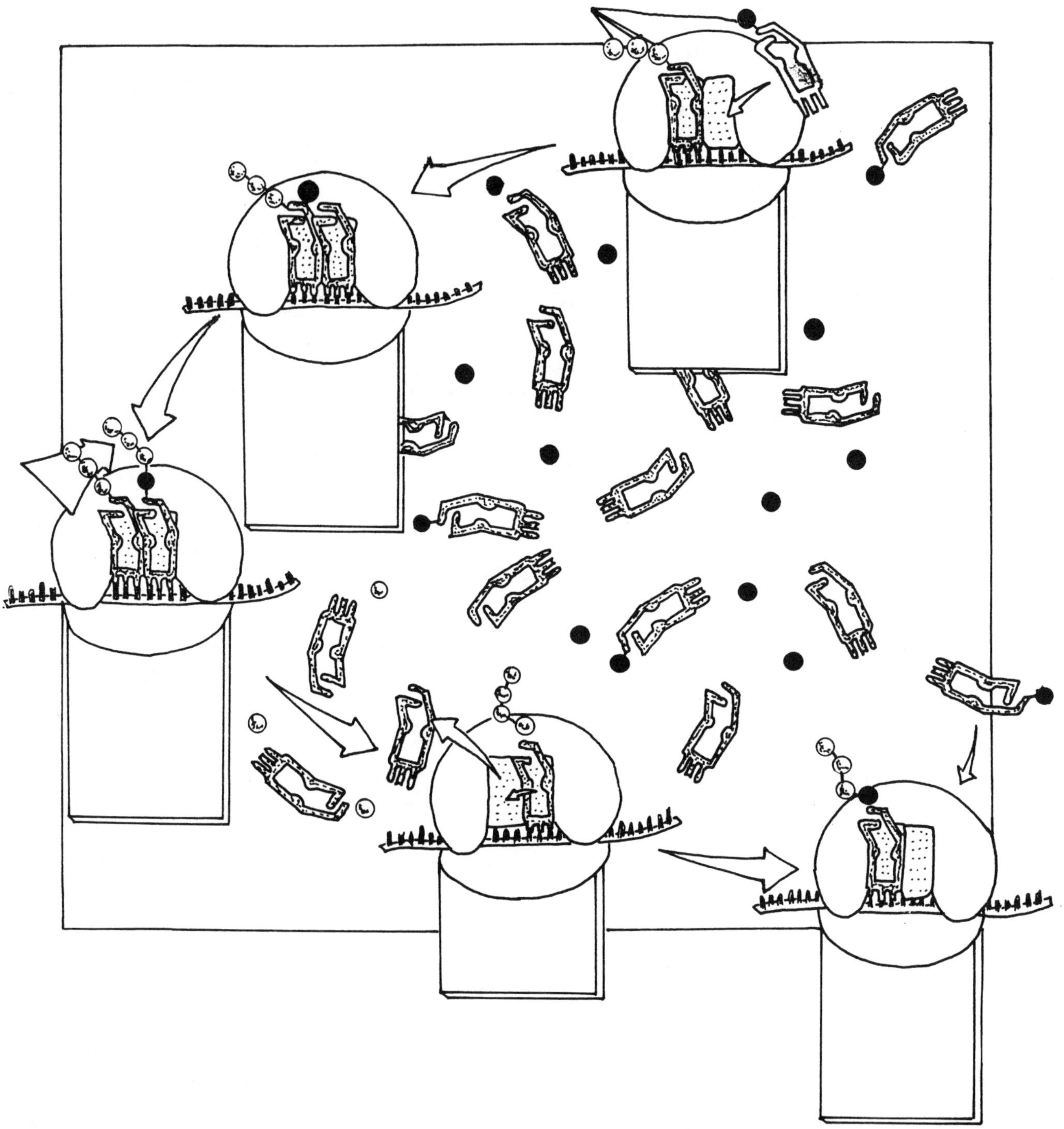

 Figure No. 16-6

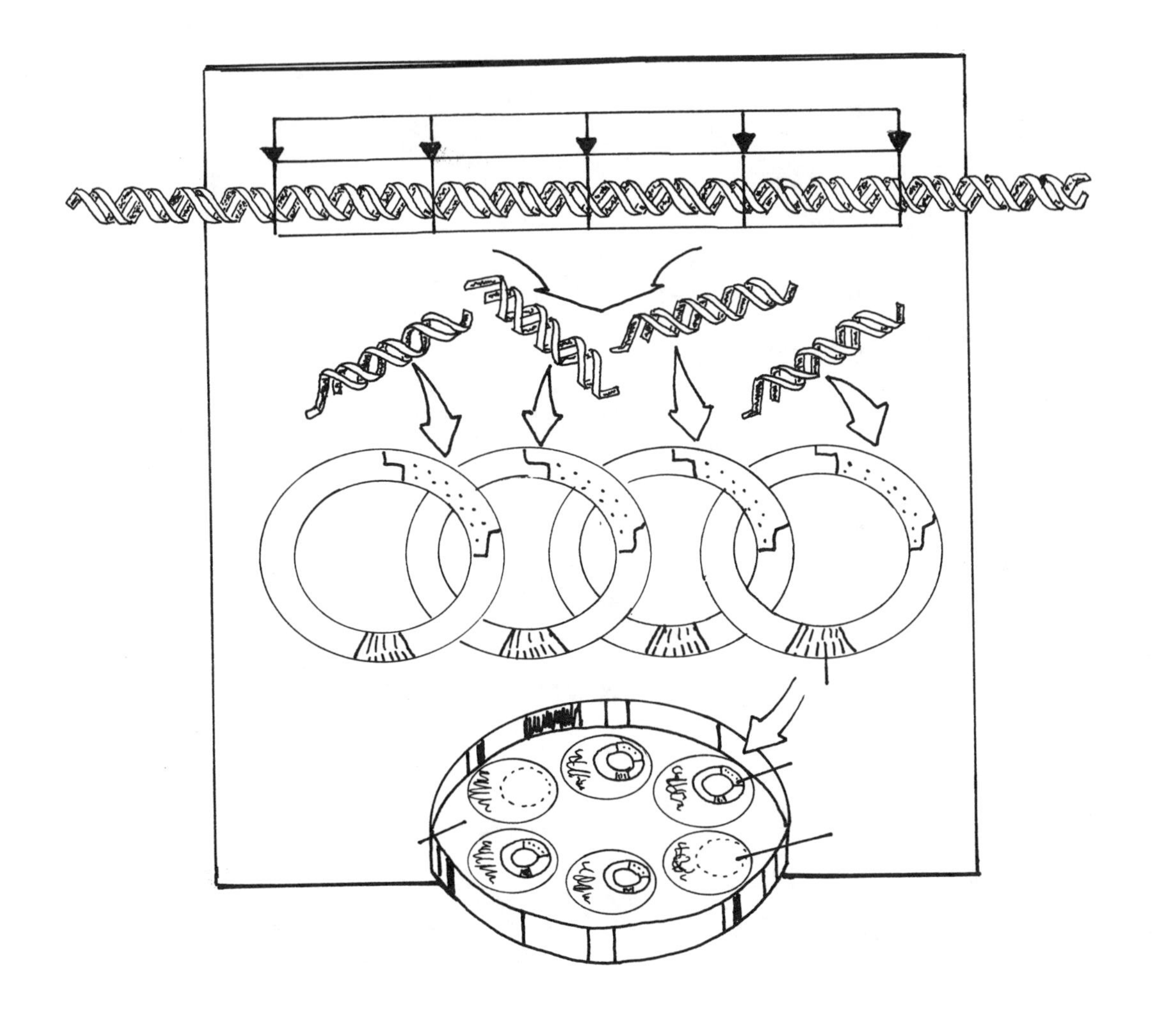

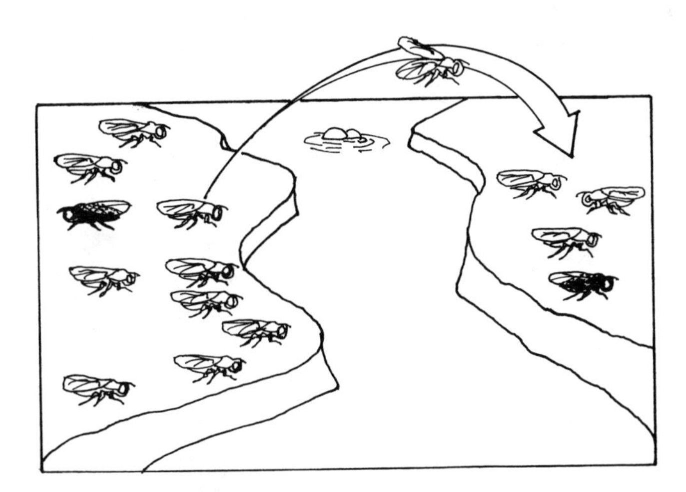

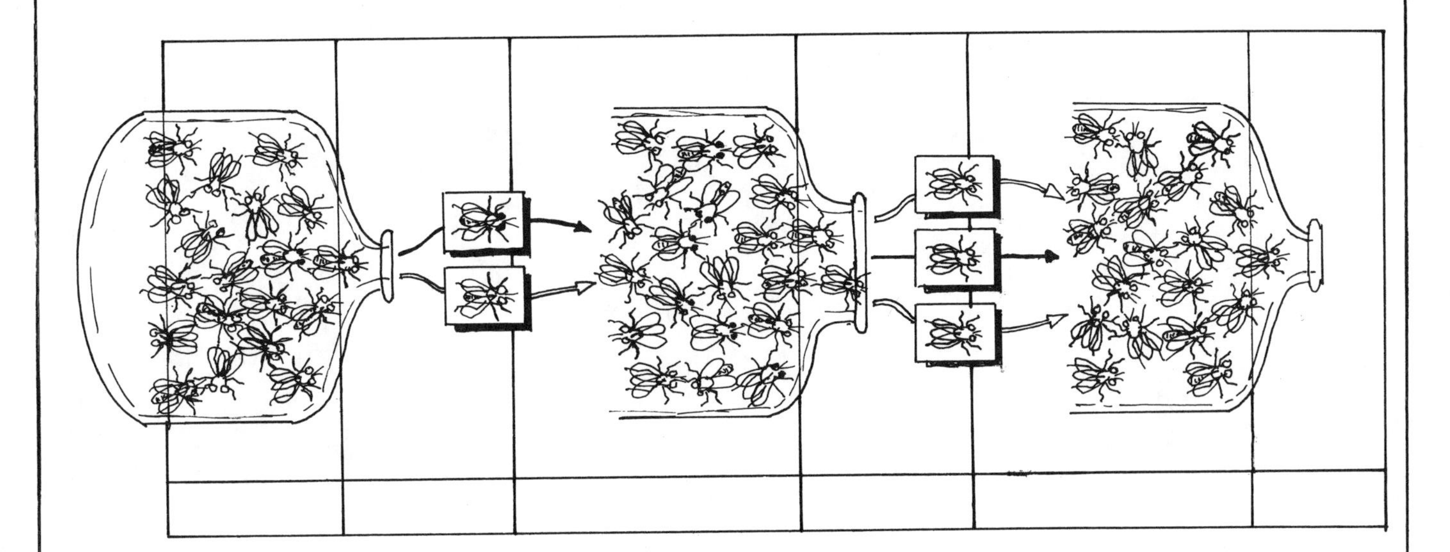

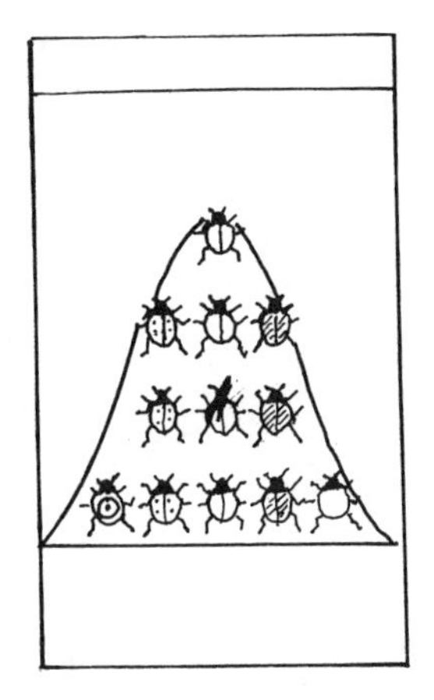

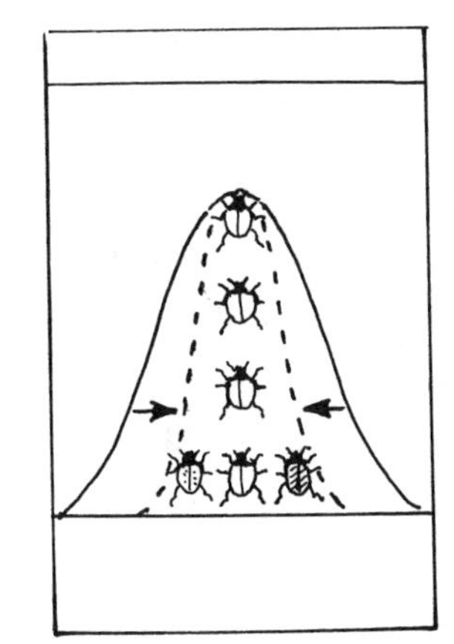

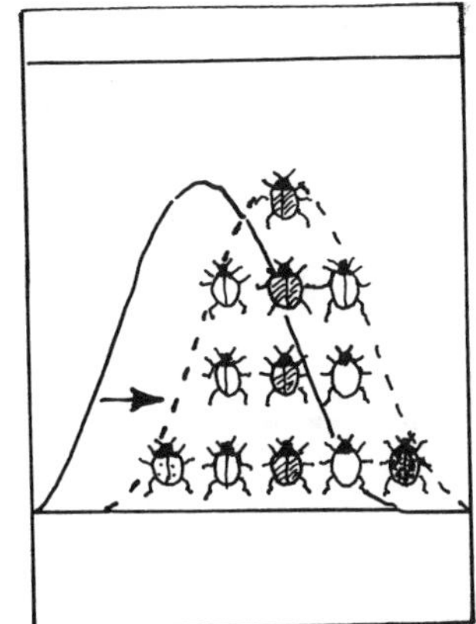

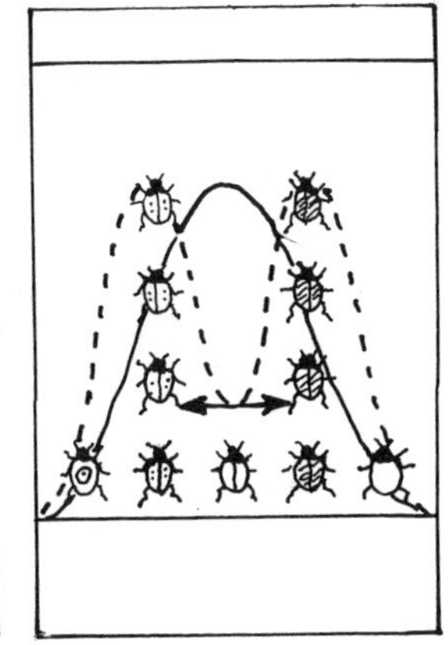

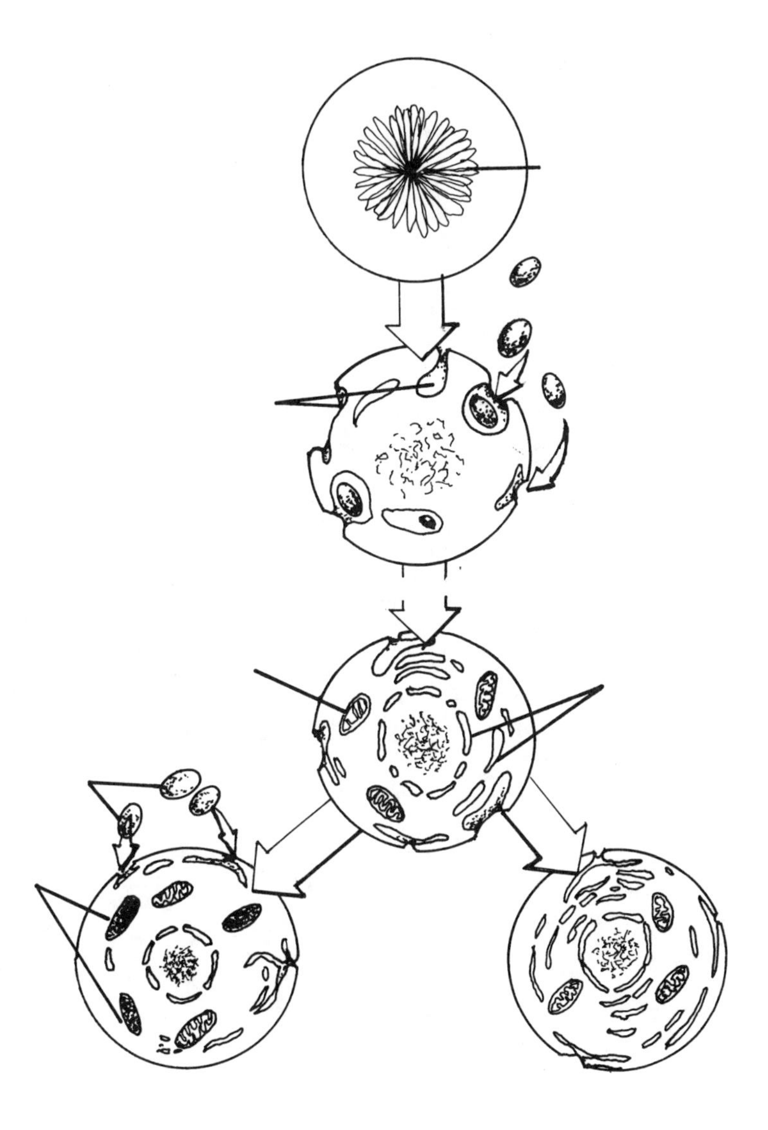

 Figure No. 20-2

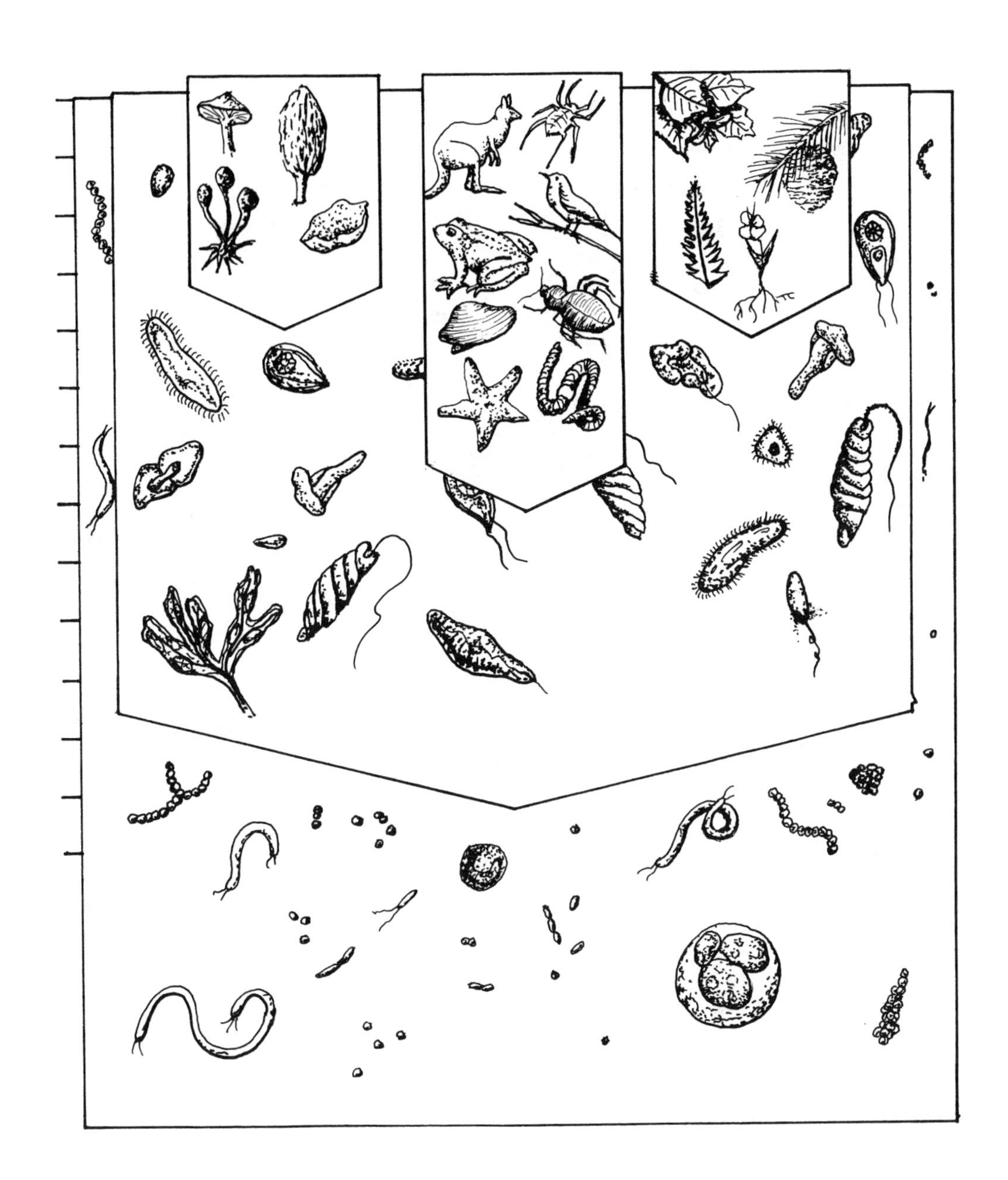

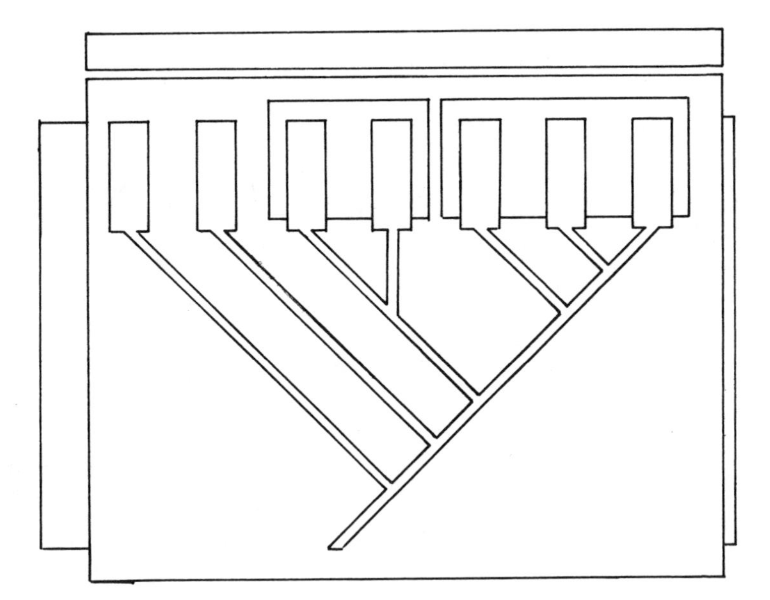

Figure No. 21-2

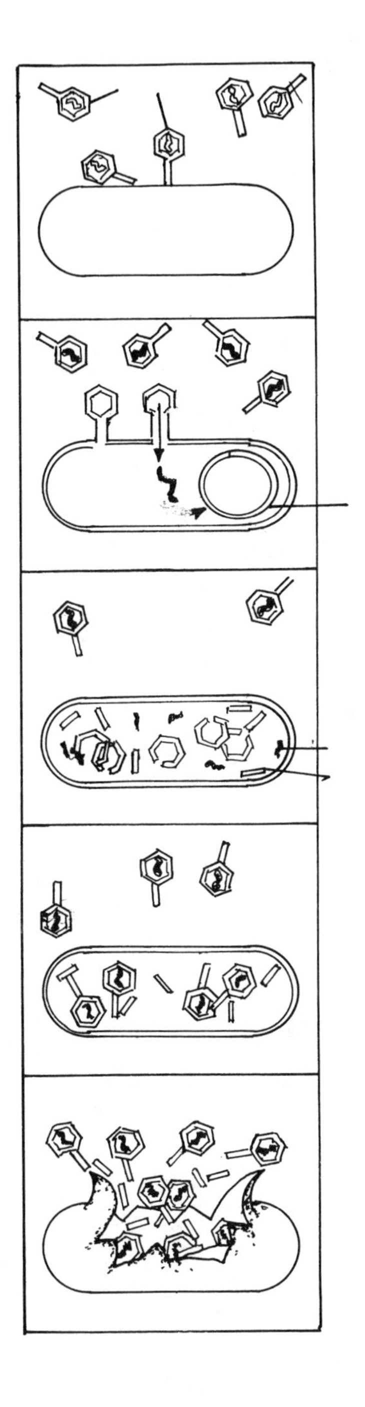

Figure No. 21-5

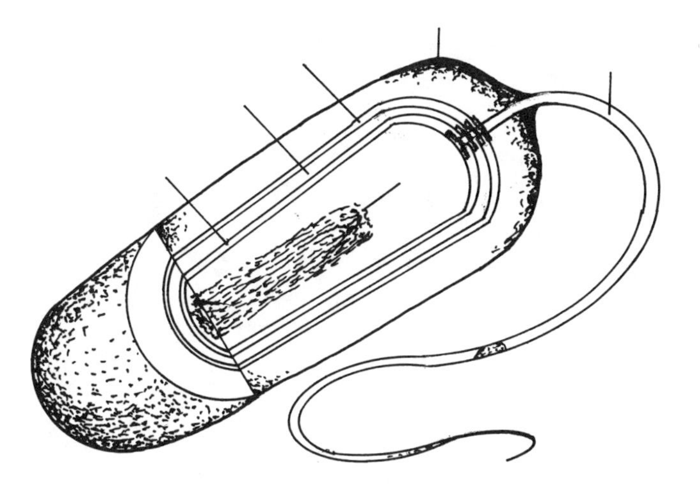

Figure No. 21-15

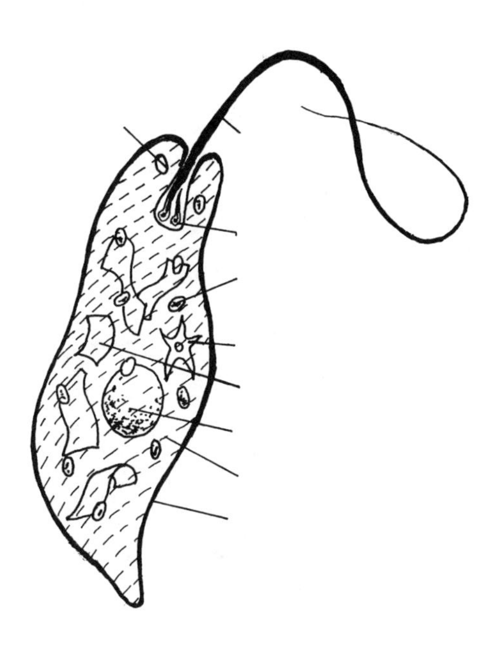

Figure No. 21-22

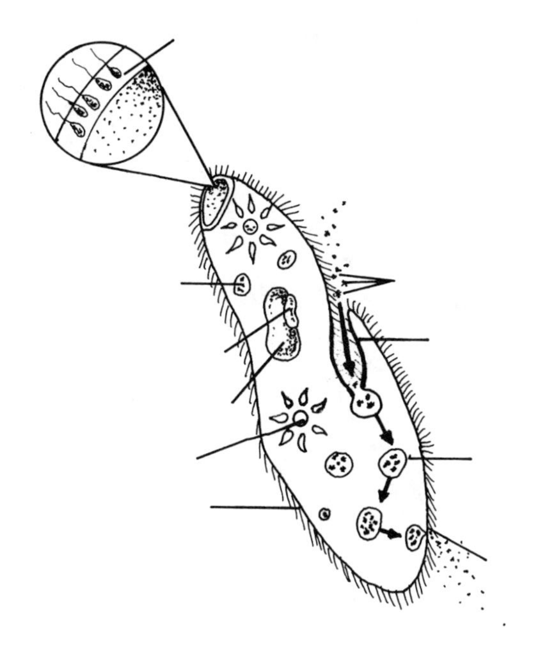

Figure No. 22-2

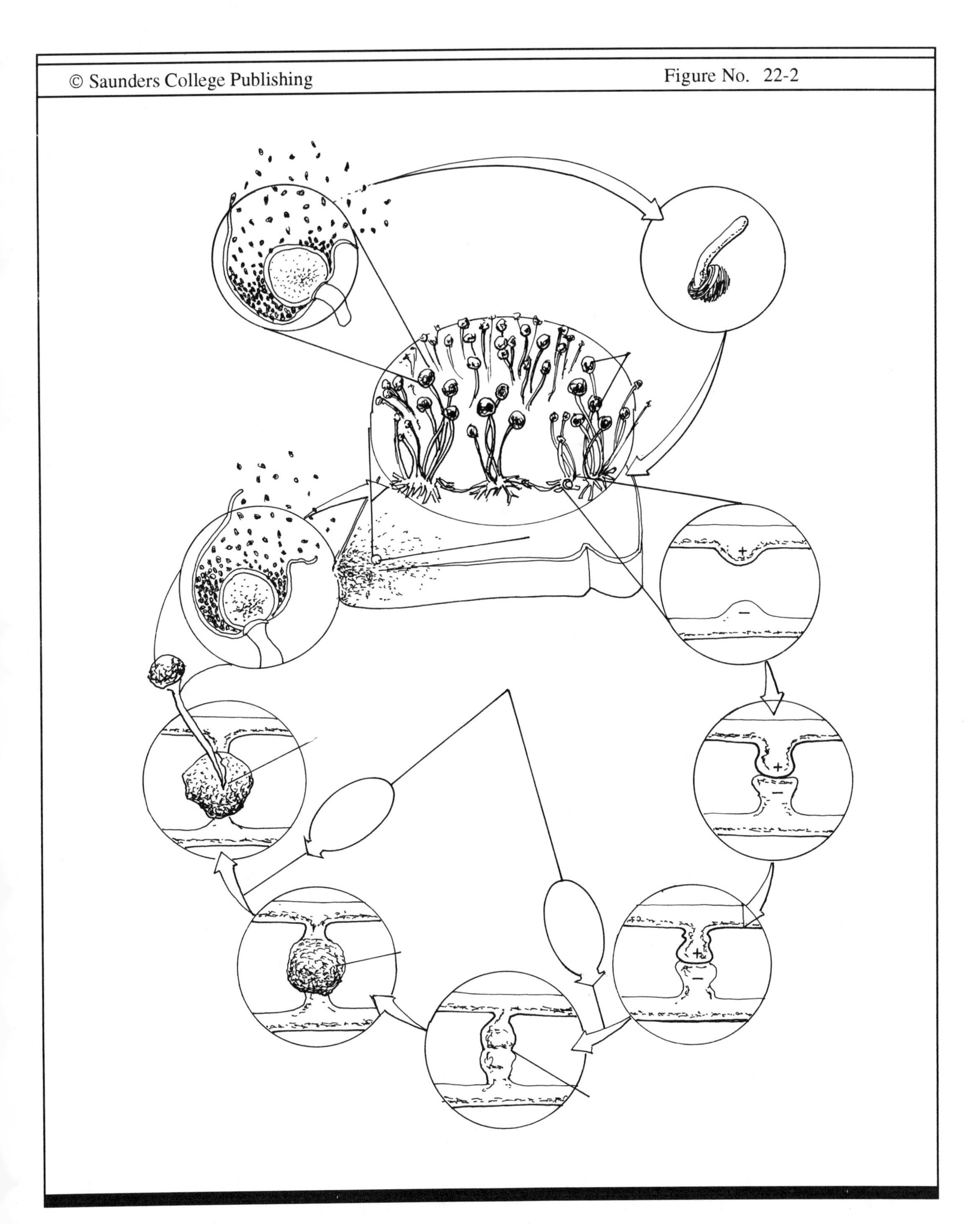

 Figure No. 22-5

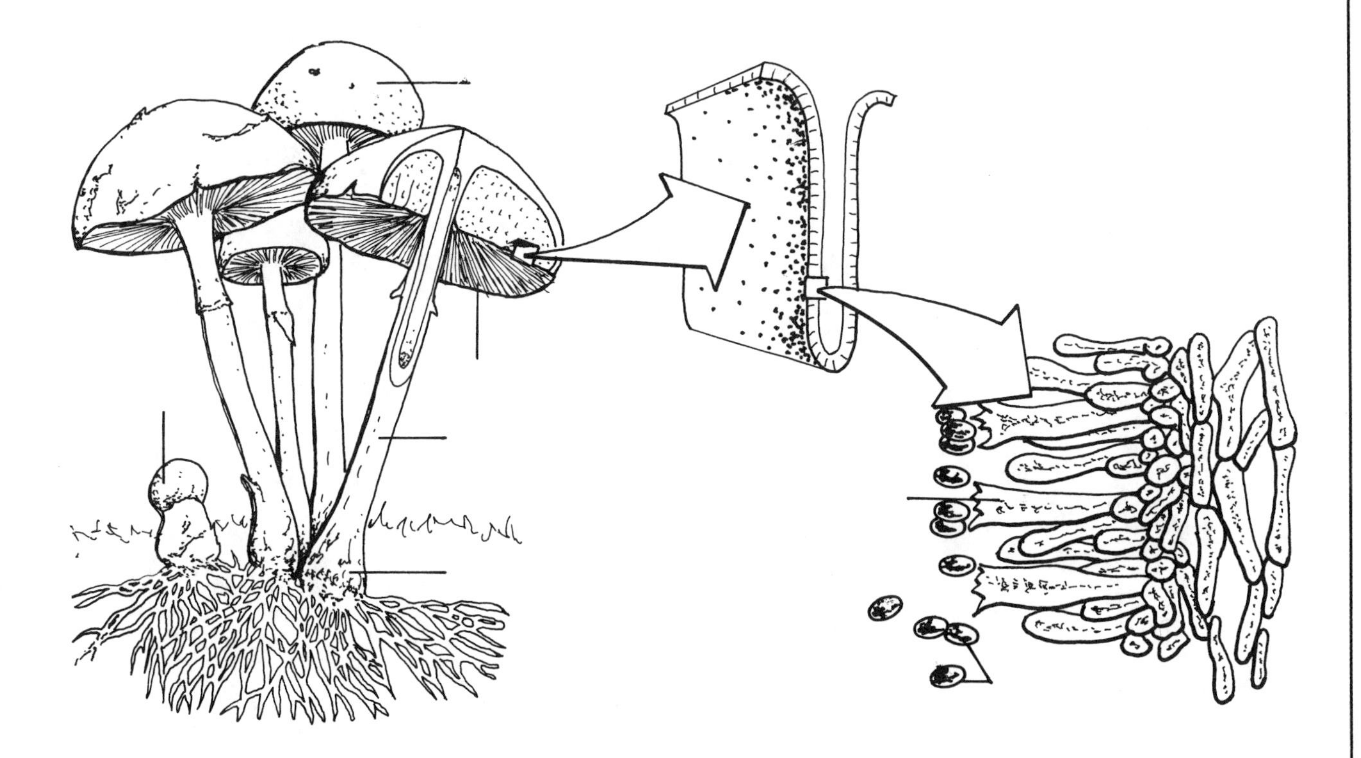

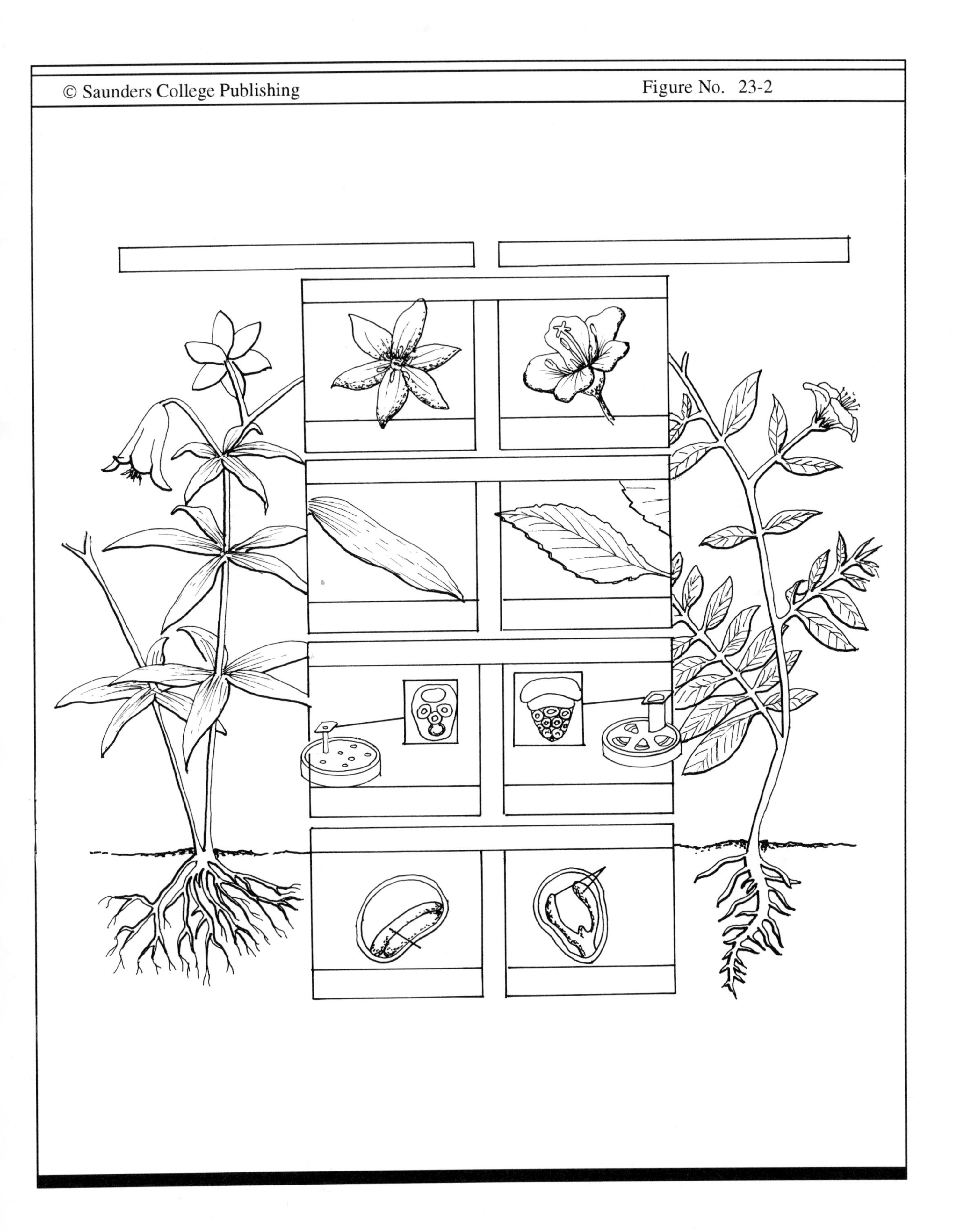
© Saunders College Publishing
Figure No. 23-2

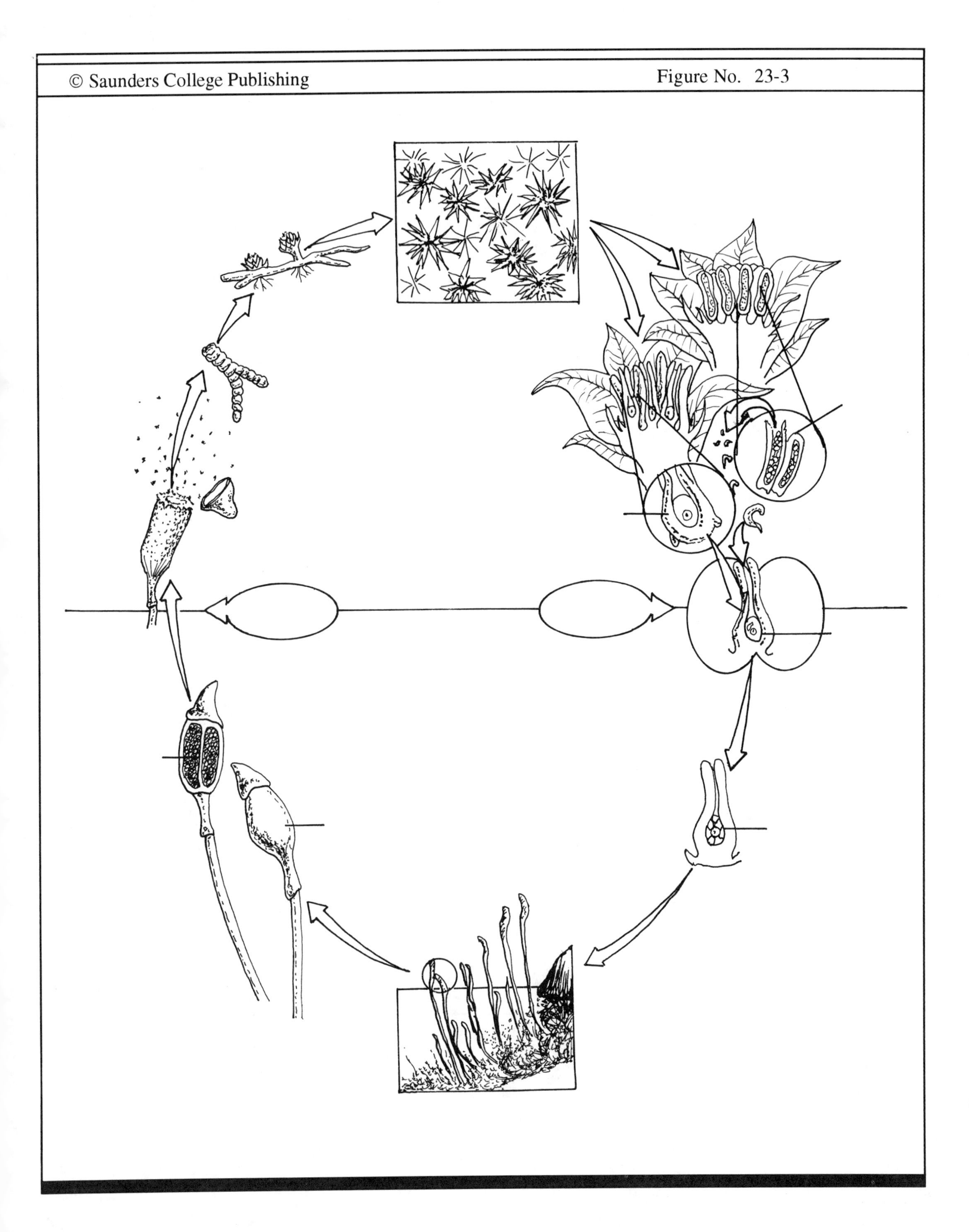
© Saunders College Publishing
Figure No. 23-3

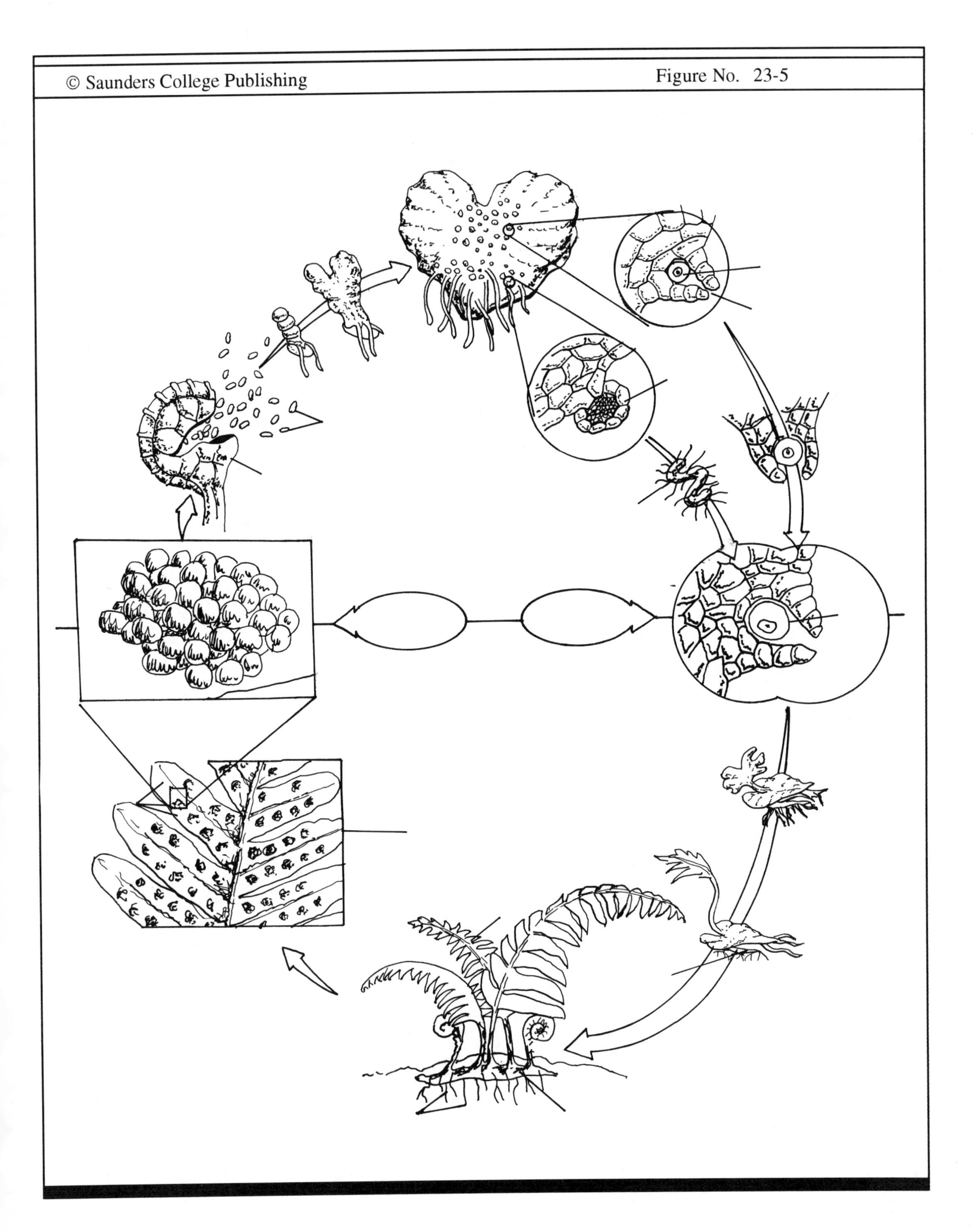
© Saunders College Publishing
Figure No. 23-5

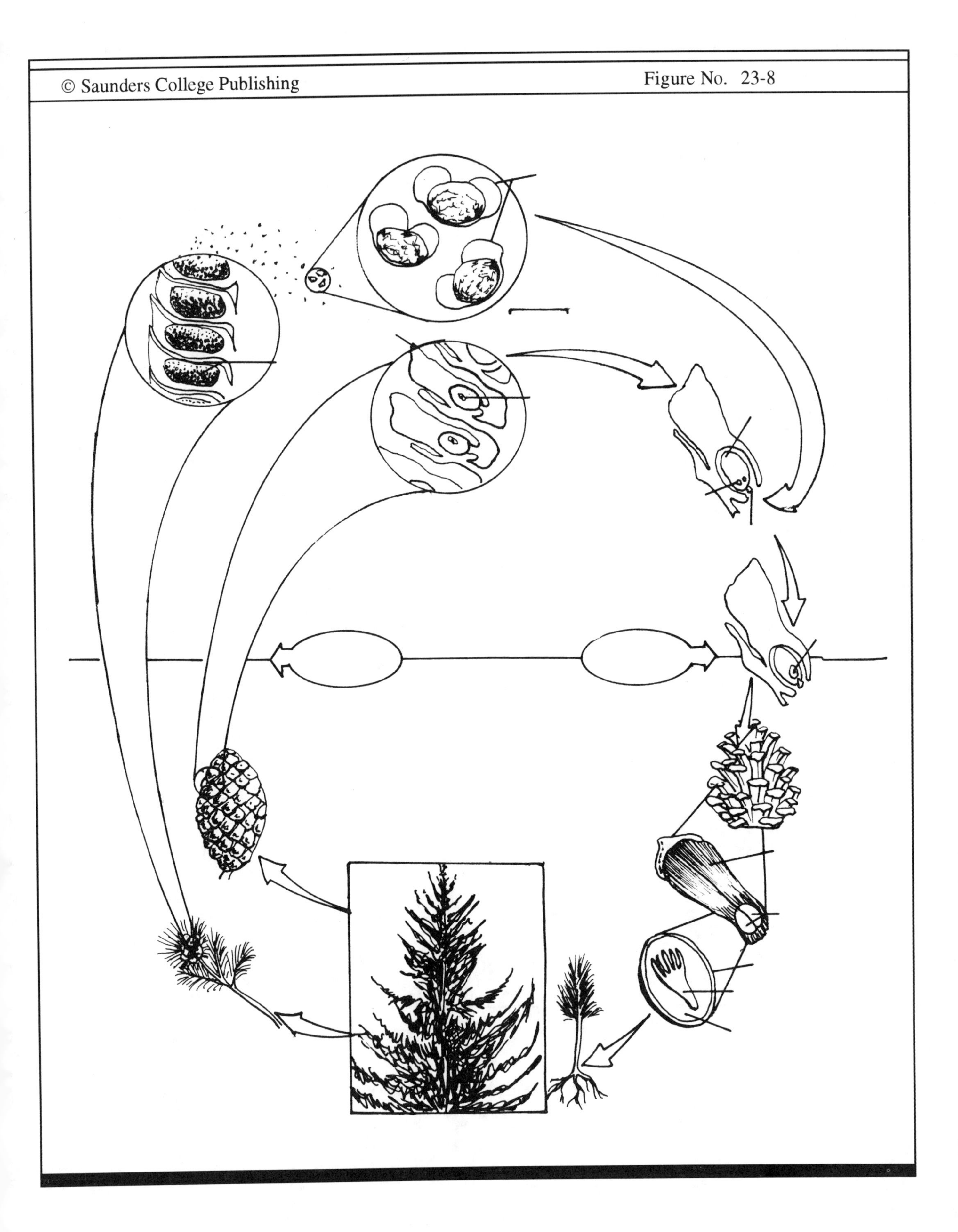
© Saunders College Publishing
Figure No. 23-8

 Figure No. 23-10

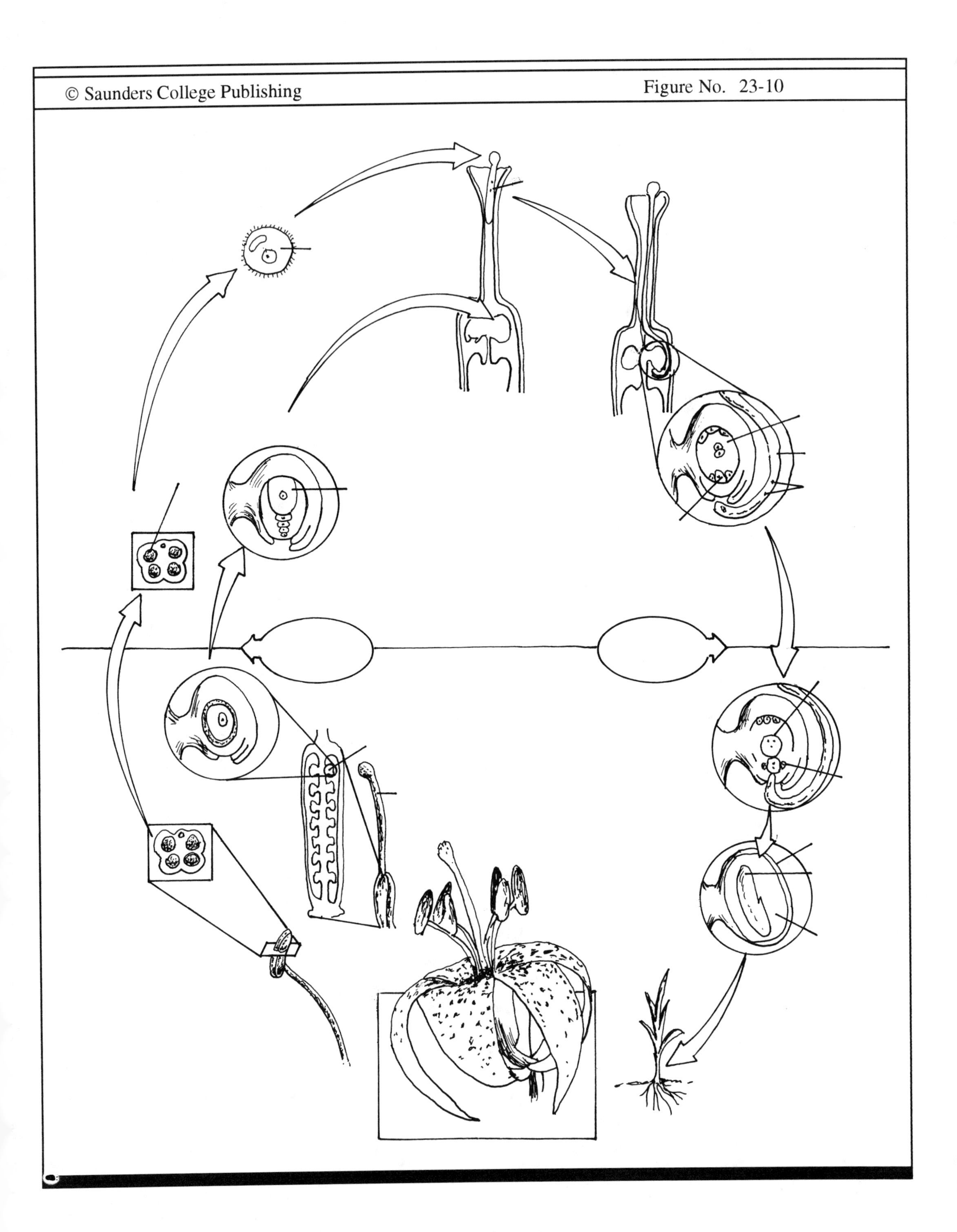

 Figure No. 24-3

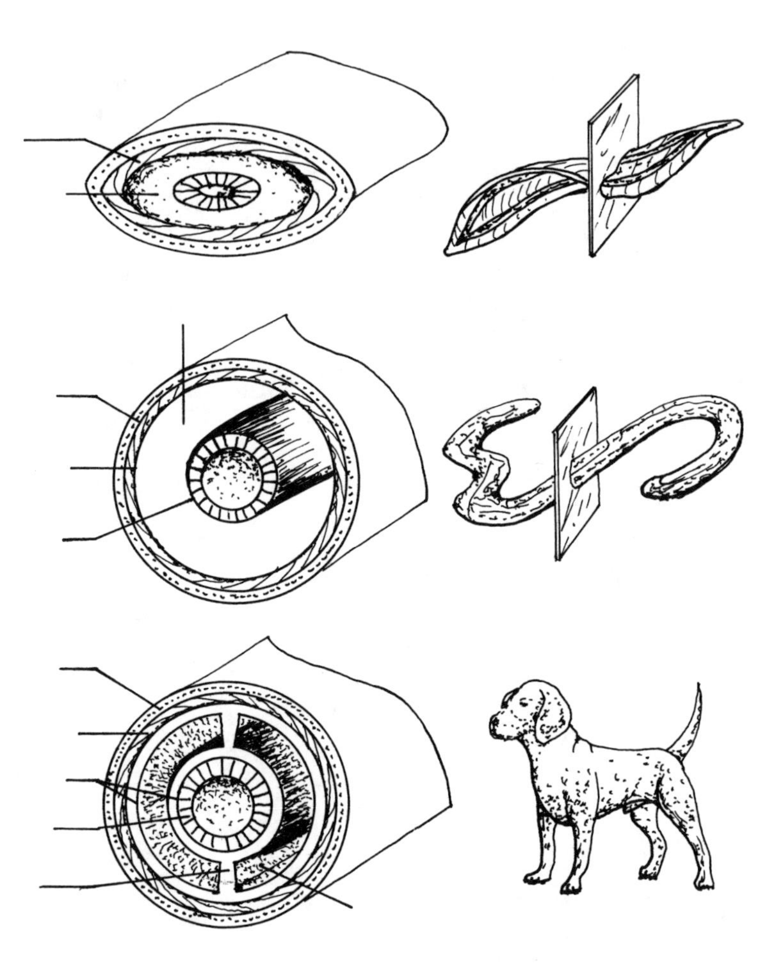

Figure No. 24-4

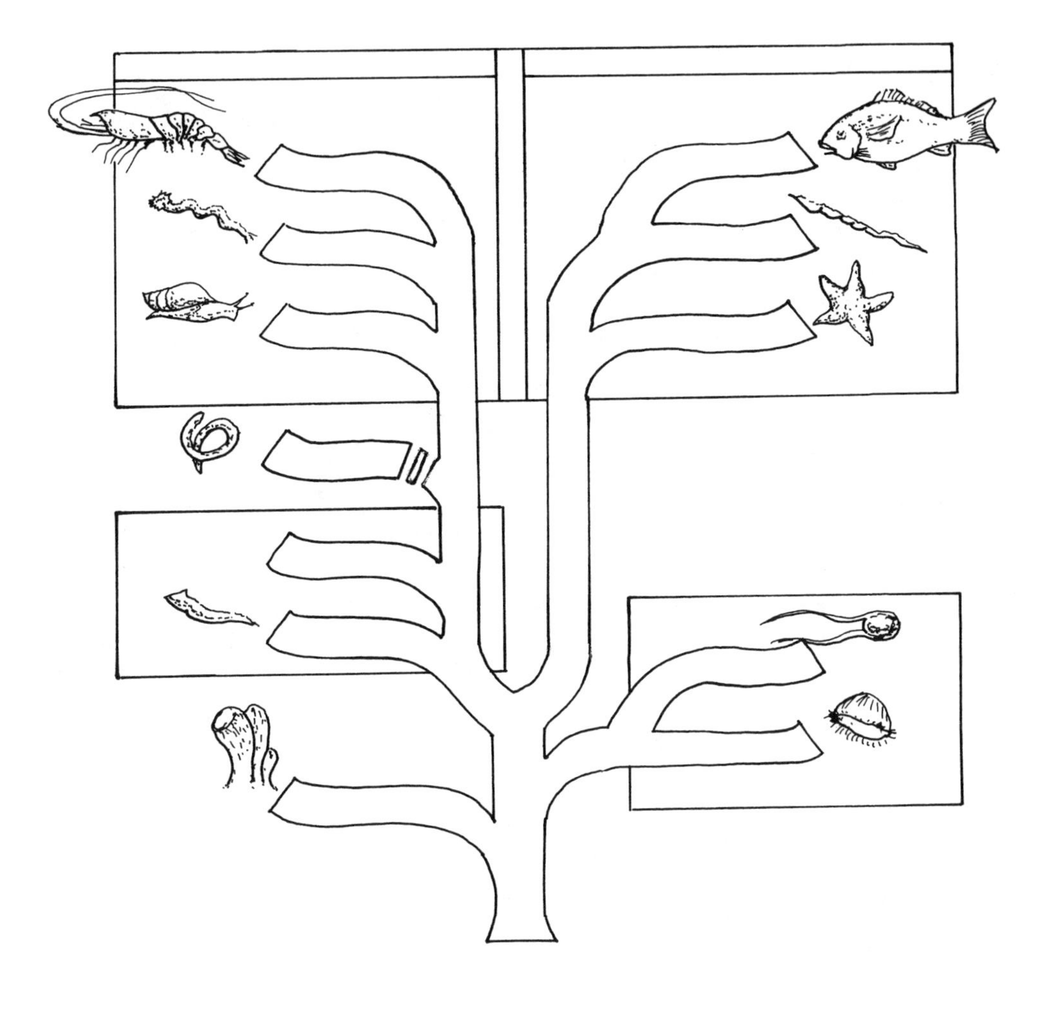

Figure No. 24-20

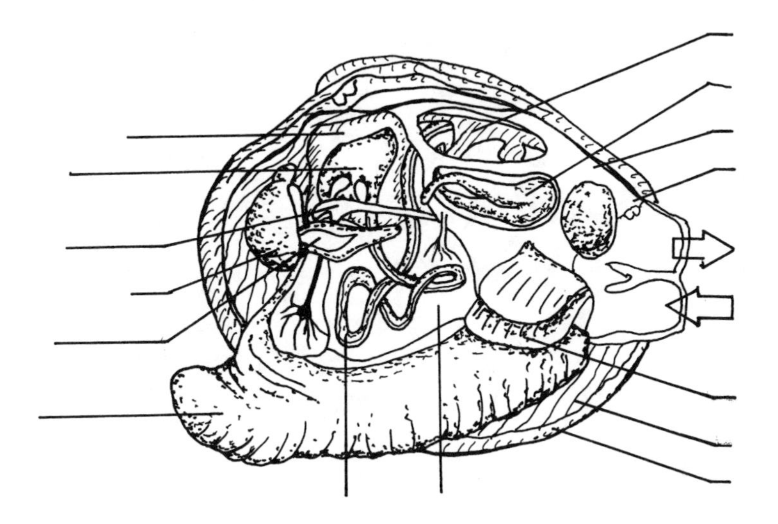

 Figure No. 24-24

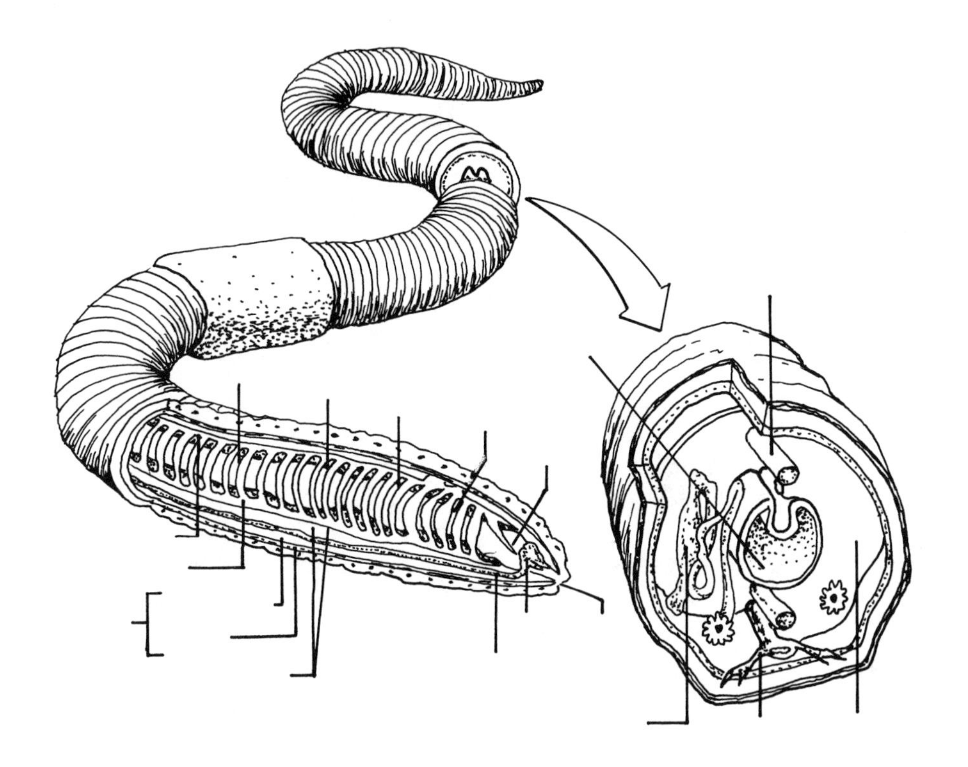

 Figure No. 24-30

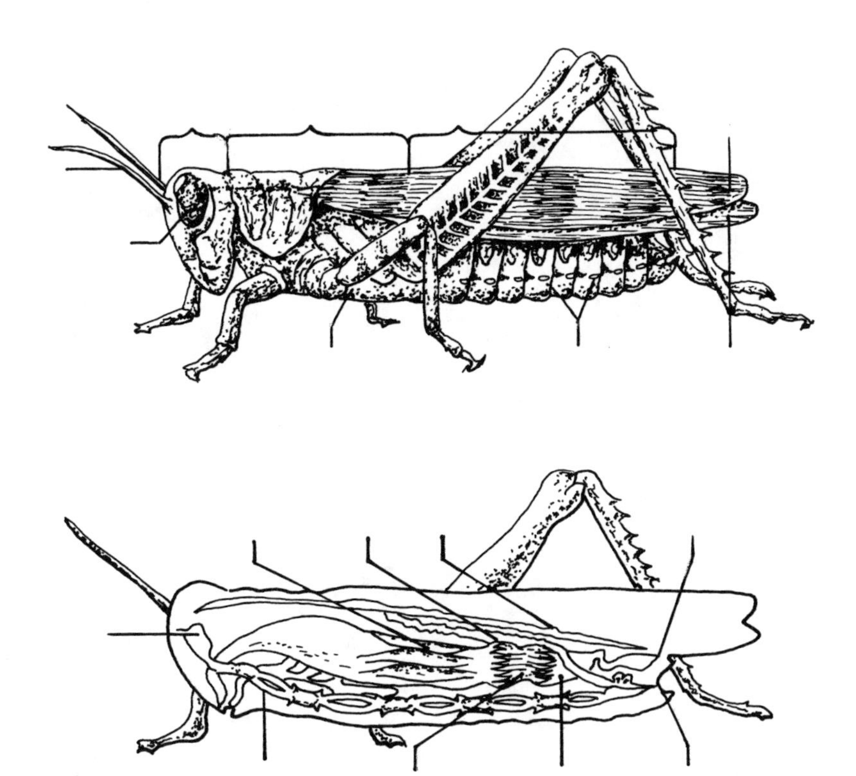

 Figure No. 25-7

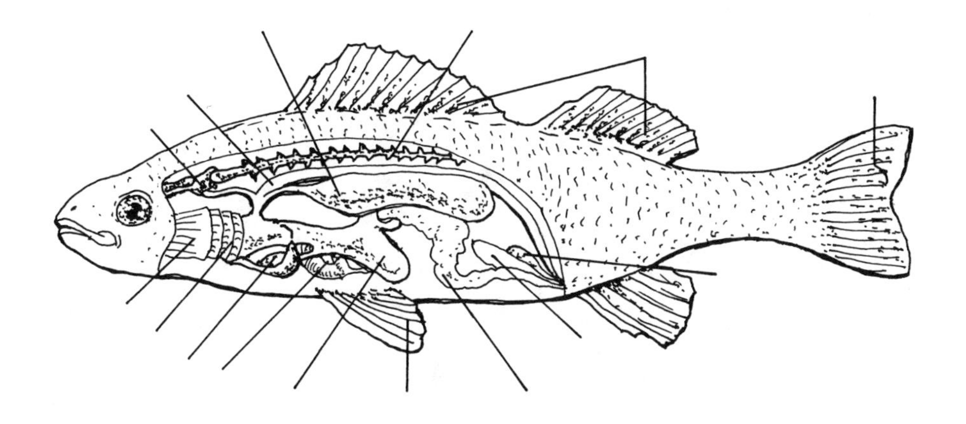

 Figure No. 26-1

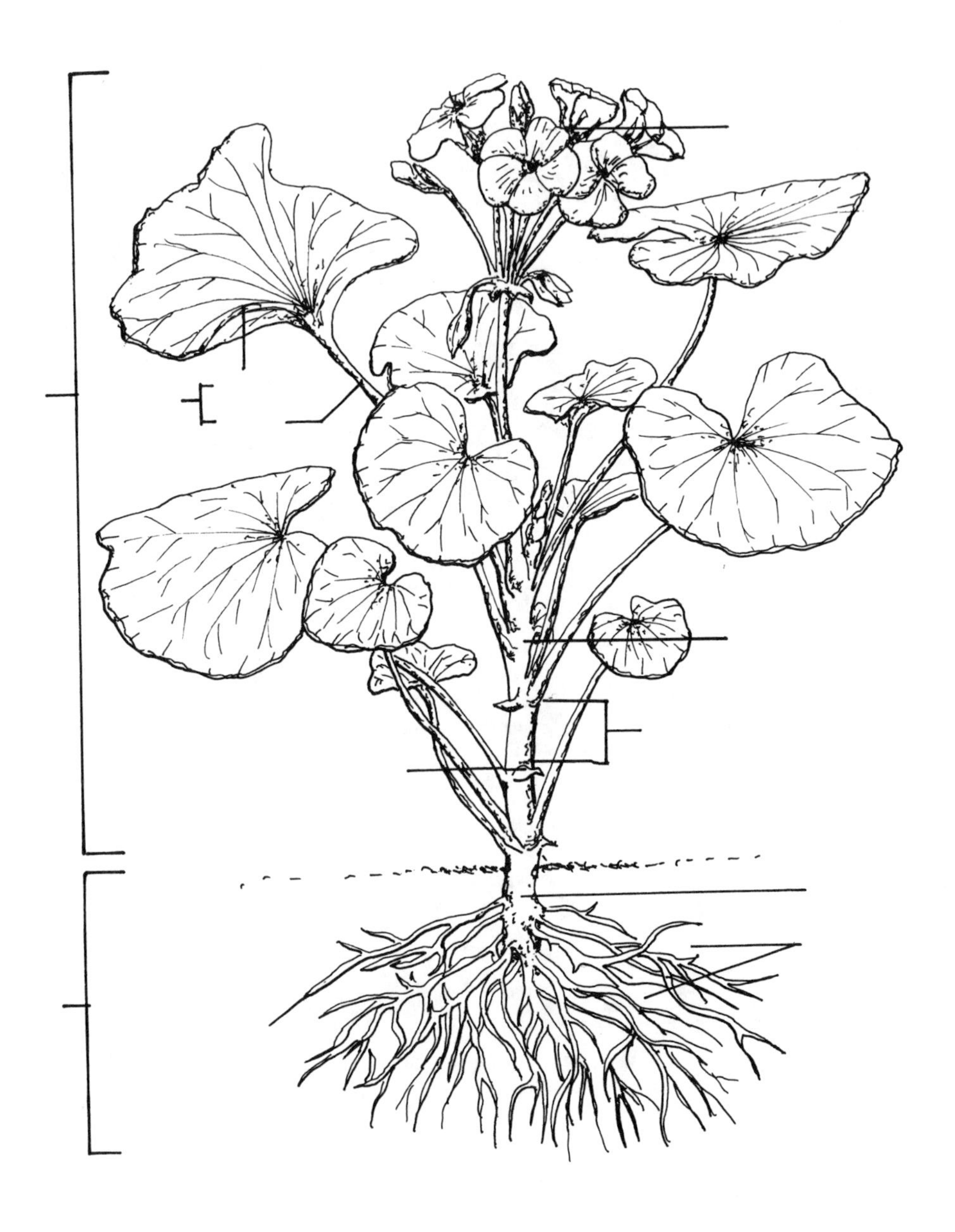

 Figure No. 26-3

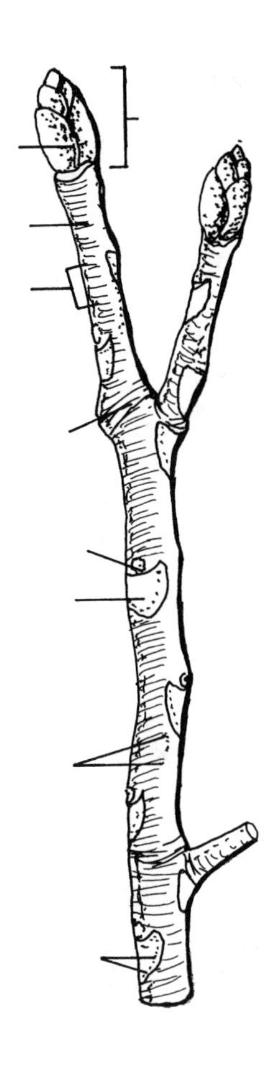

 Figure No. 26-14

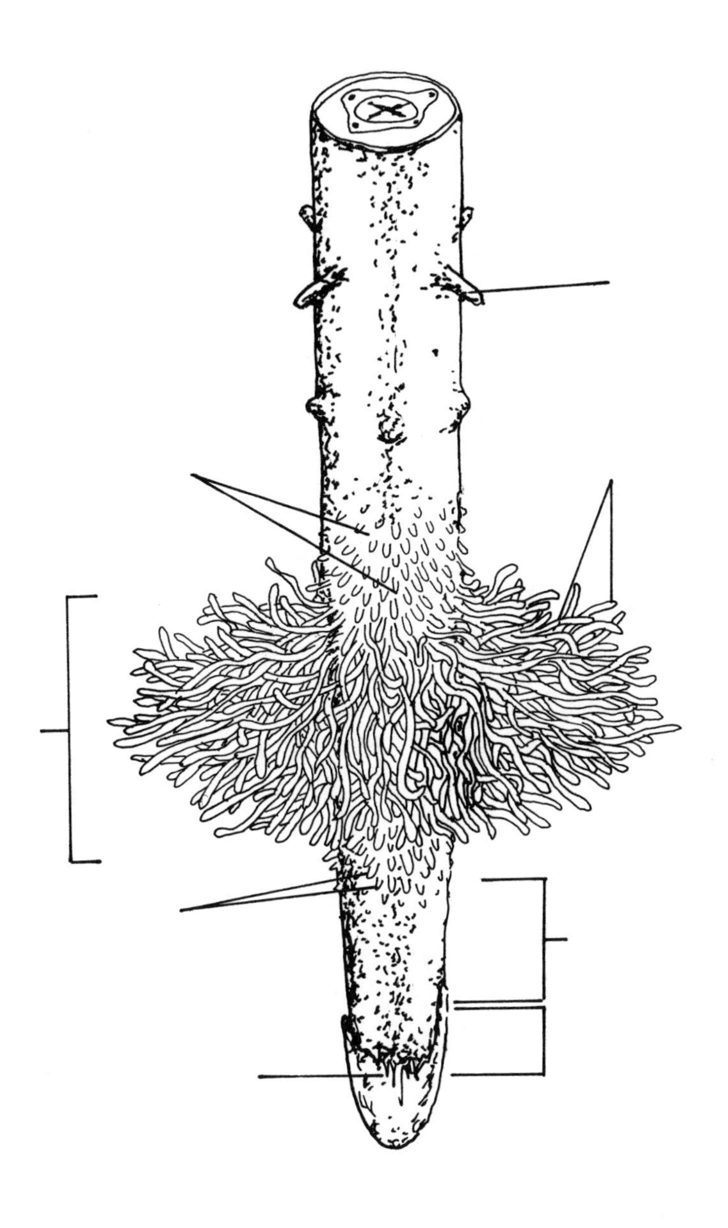

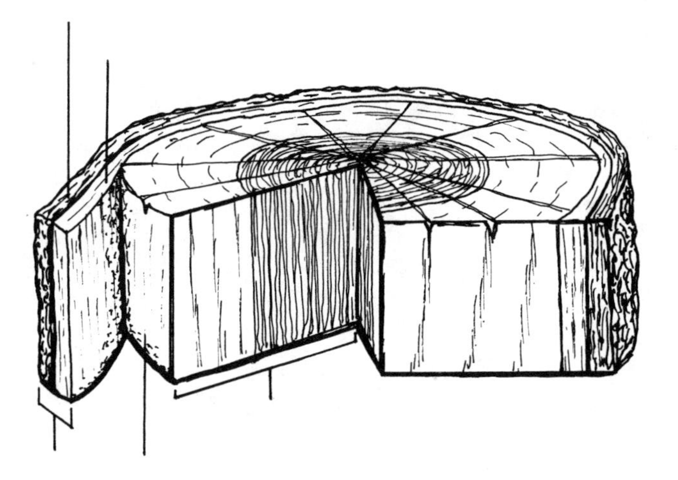

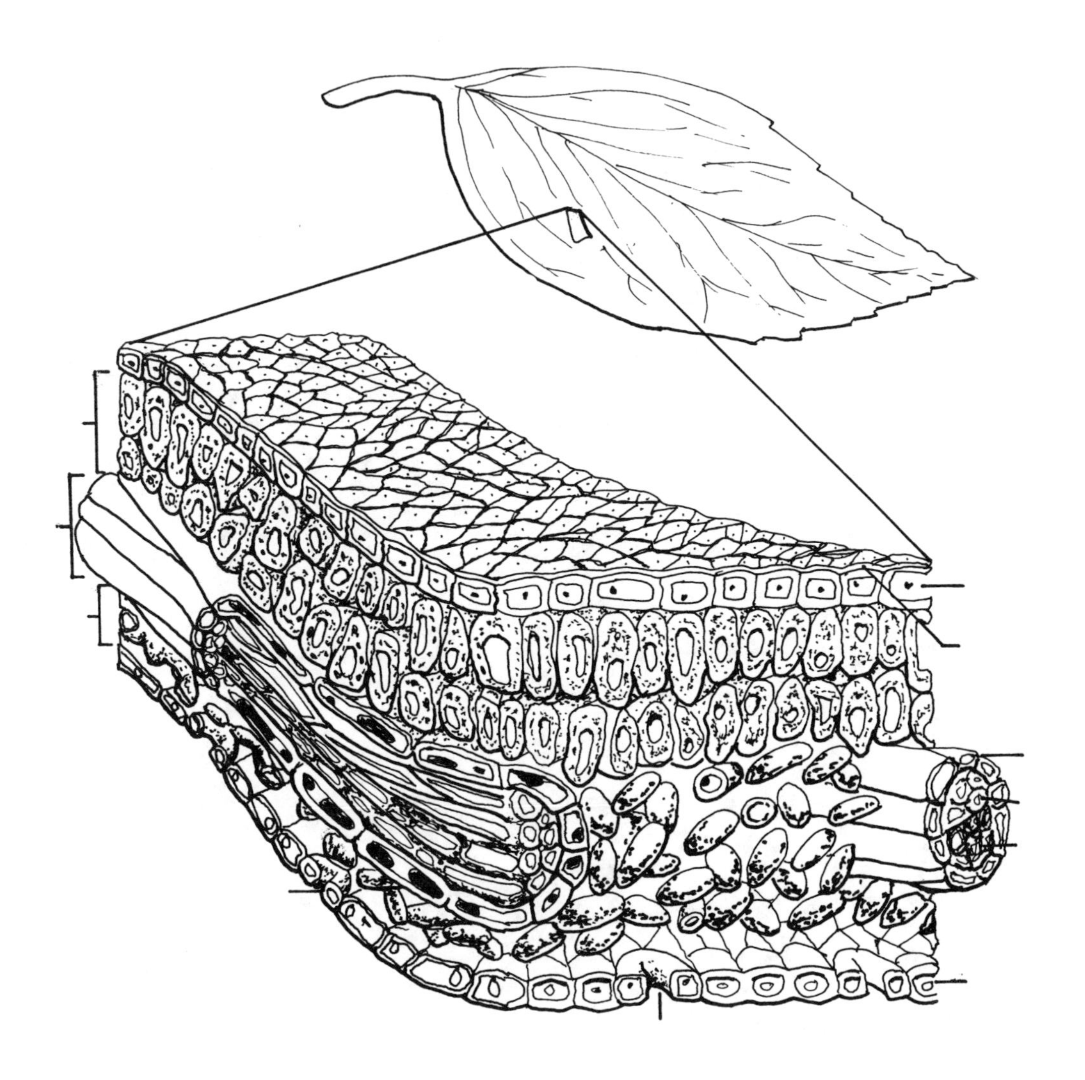

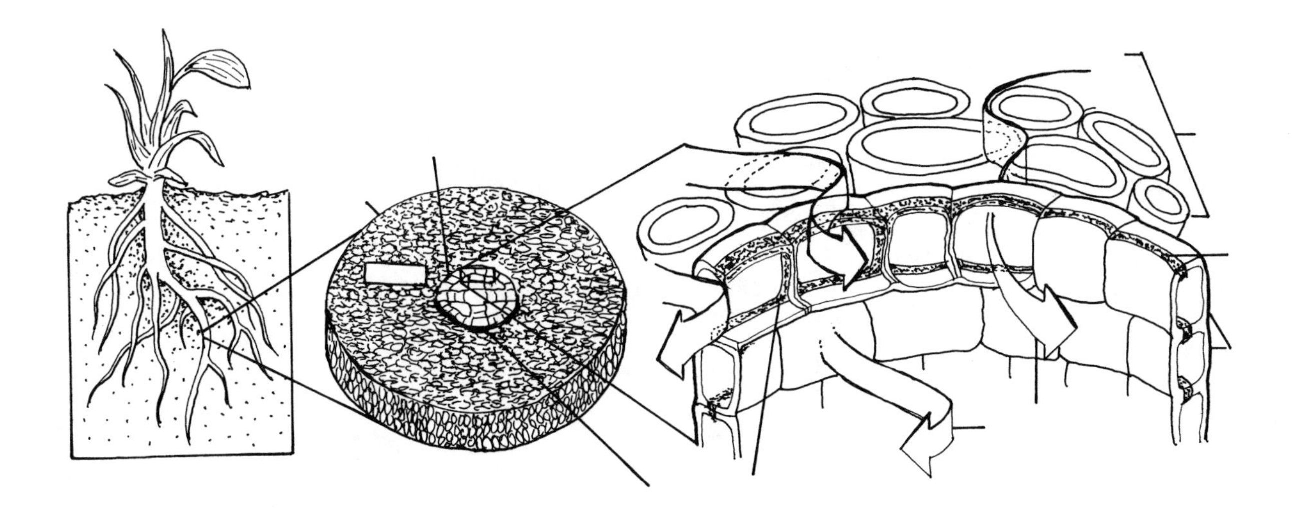

 Figure No. 28-12

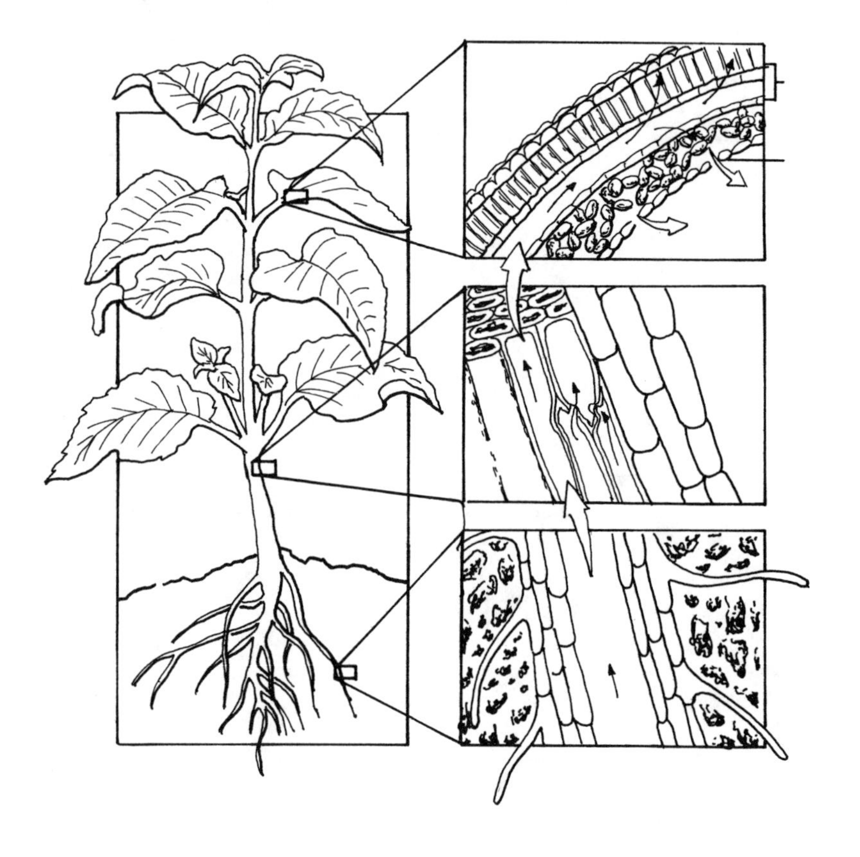

 Figure No. 28-13

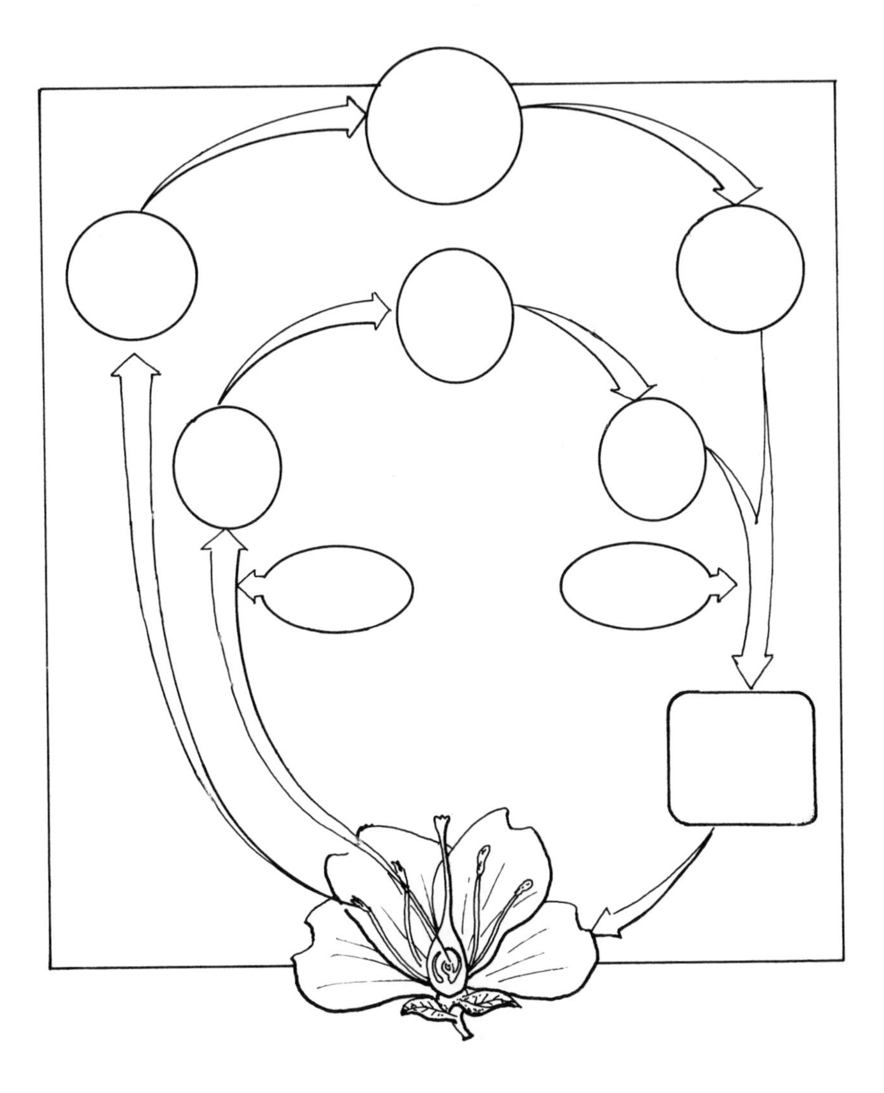

Figure No. 29-14

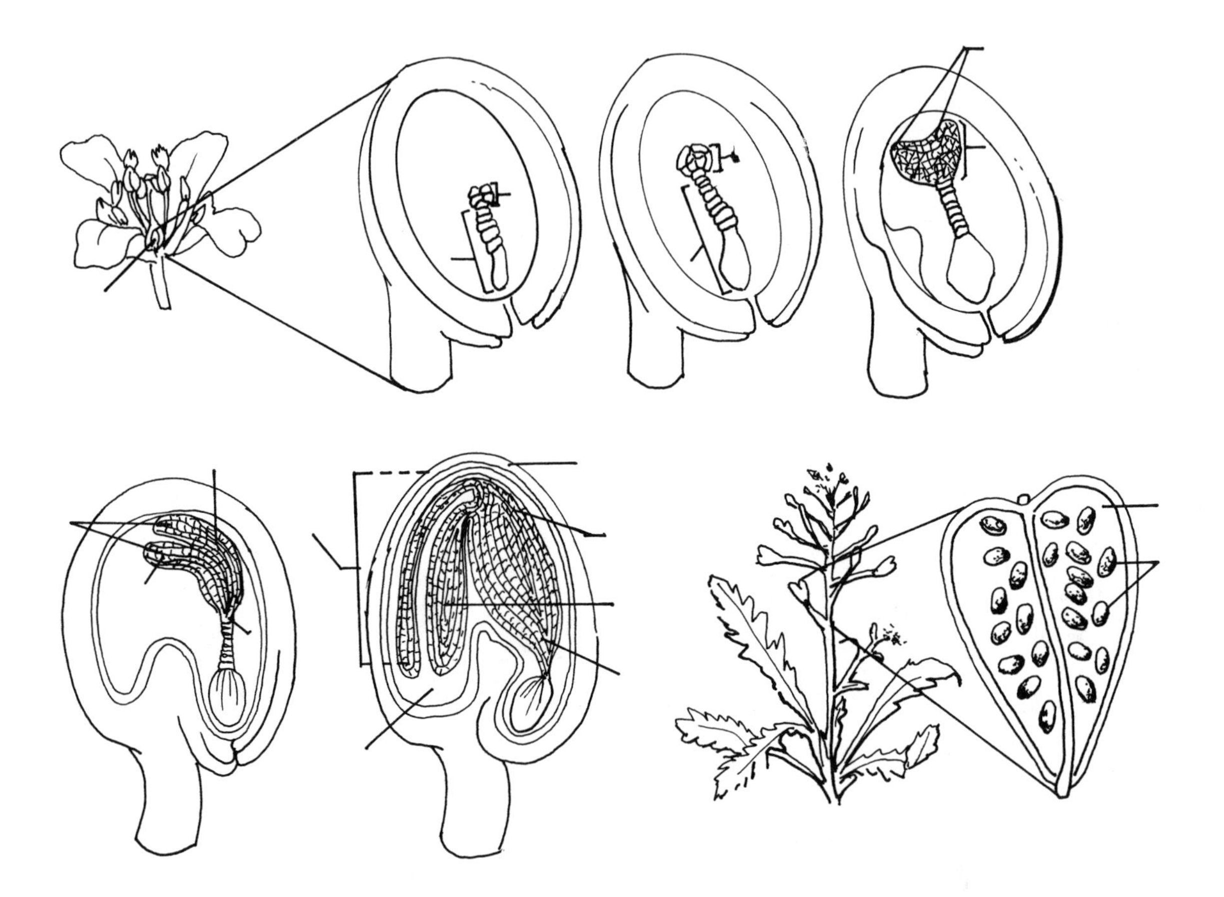

Figure No. 30-2

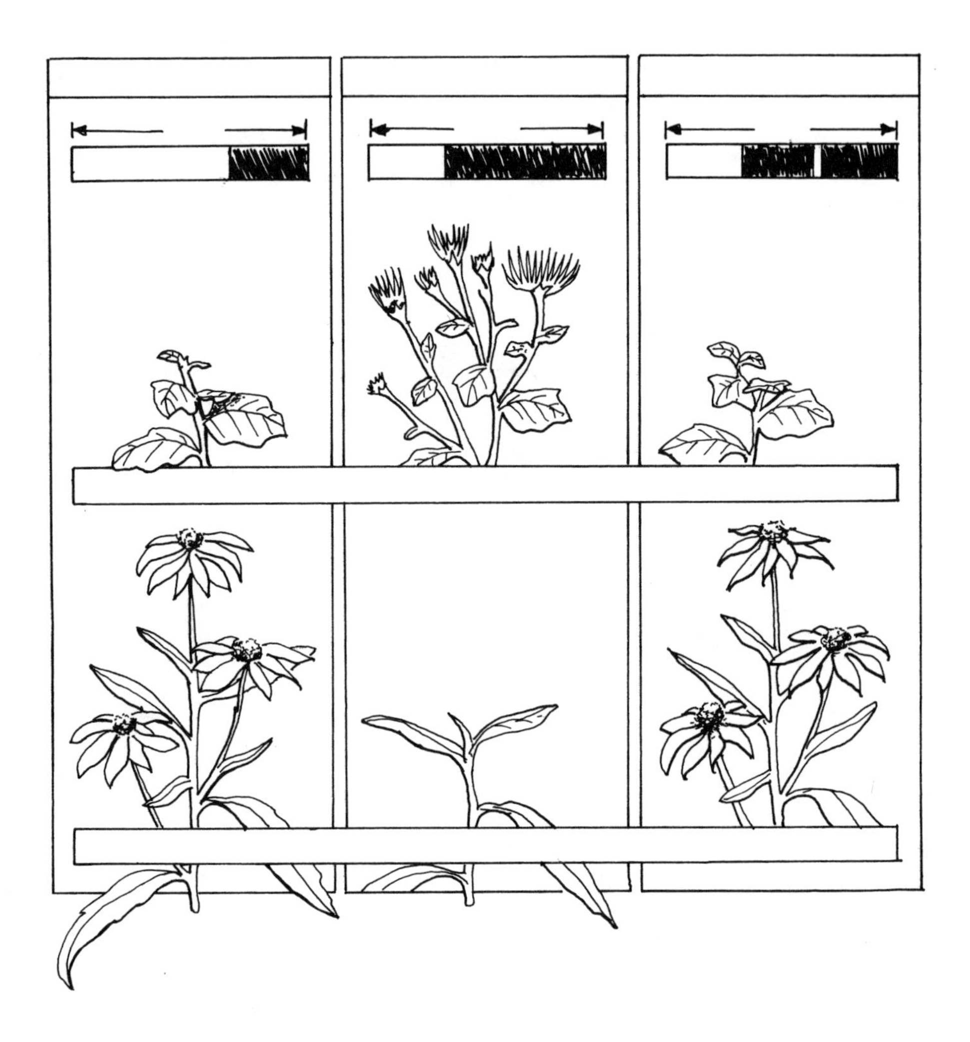

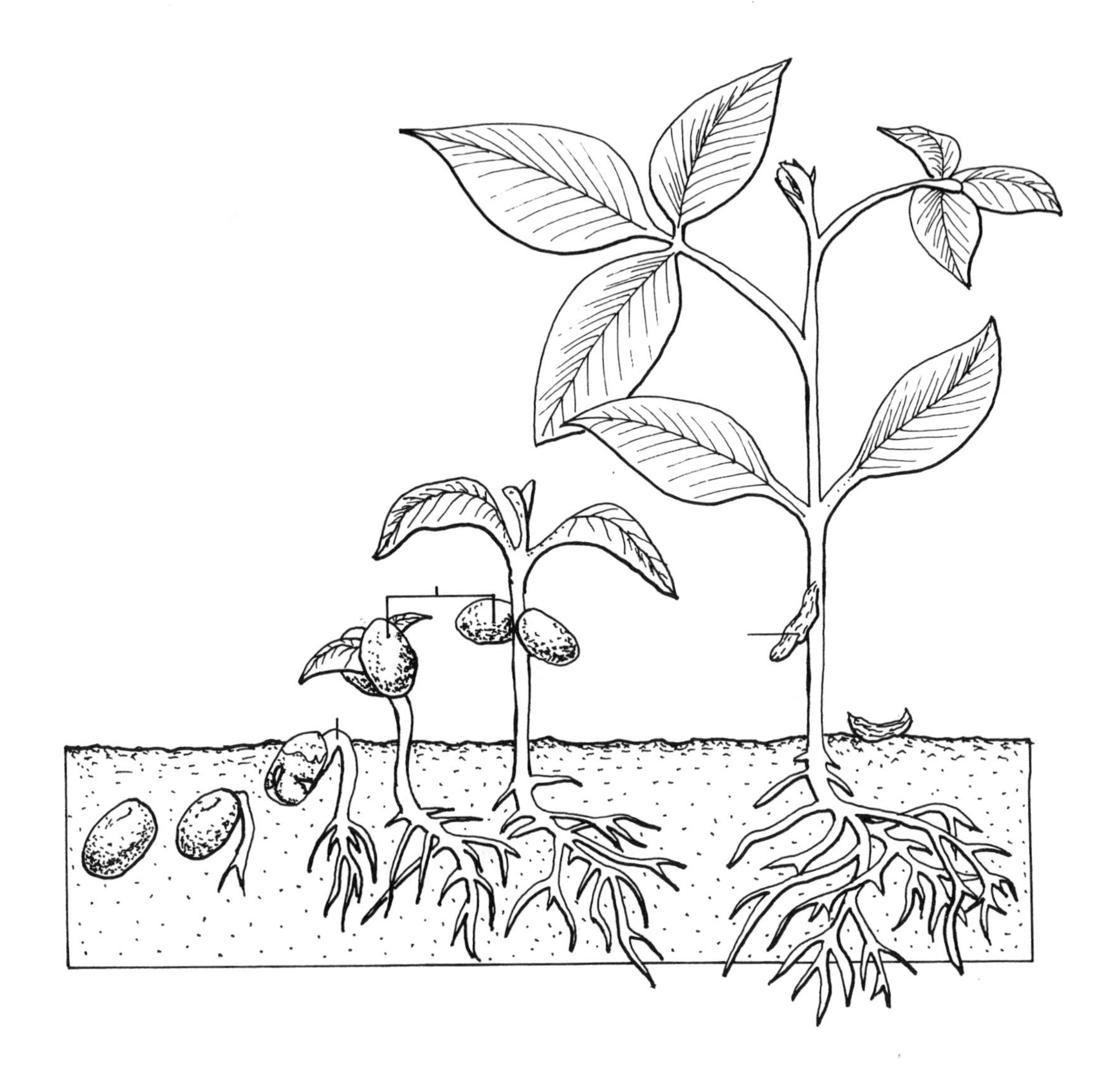

 Figure No. 30-4

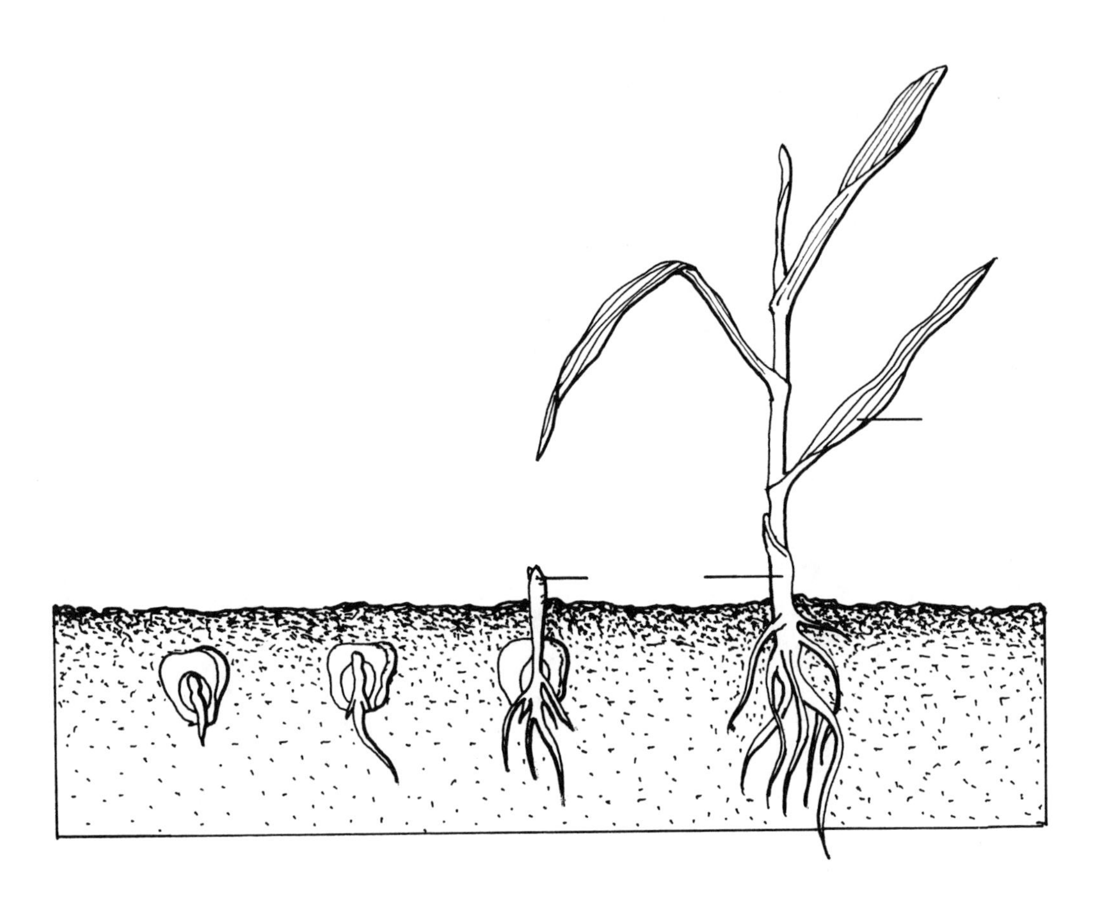

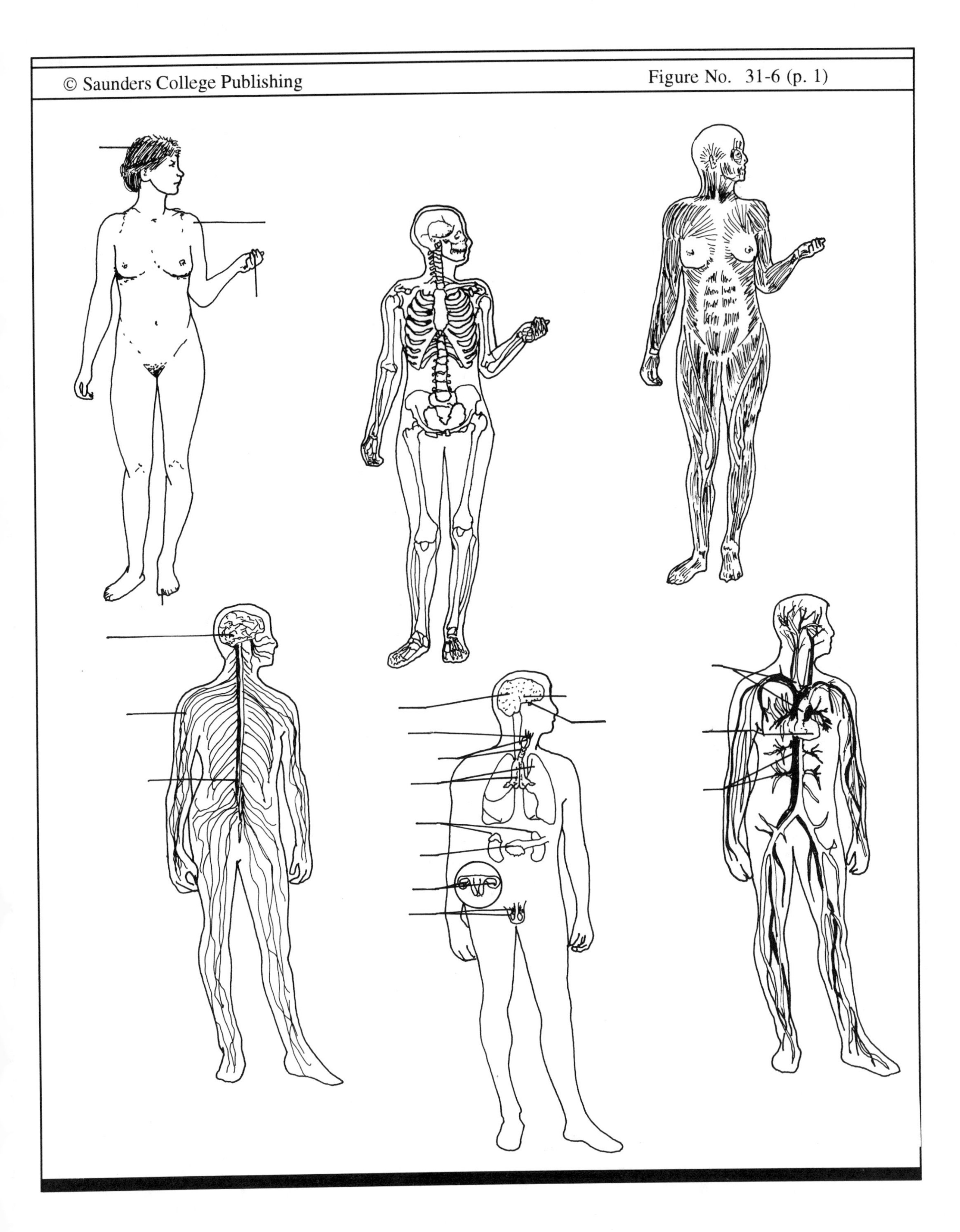
© Saunders College Publishing
Figure No. 31-6 (p. 1)

 Figure No. 31-6 (p. 2)

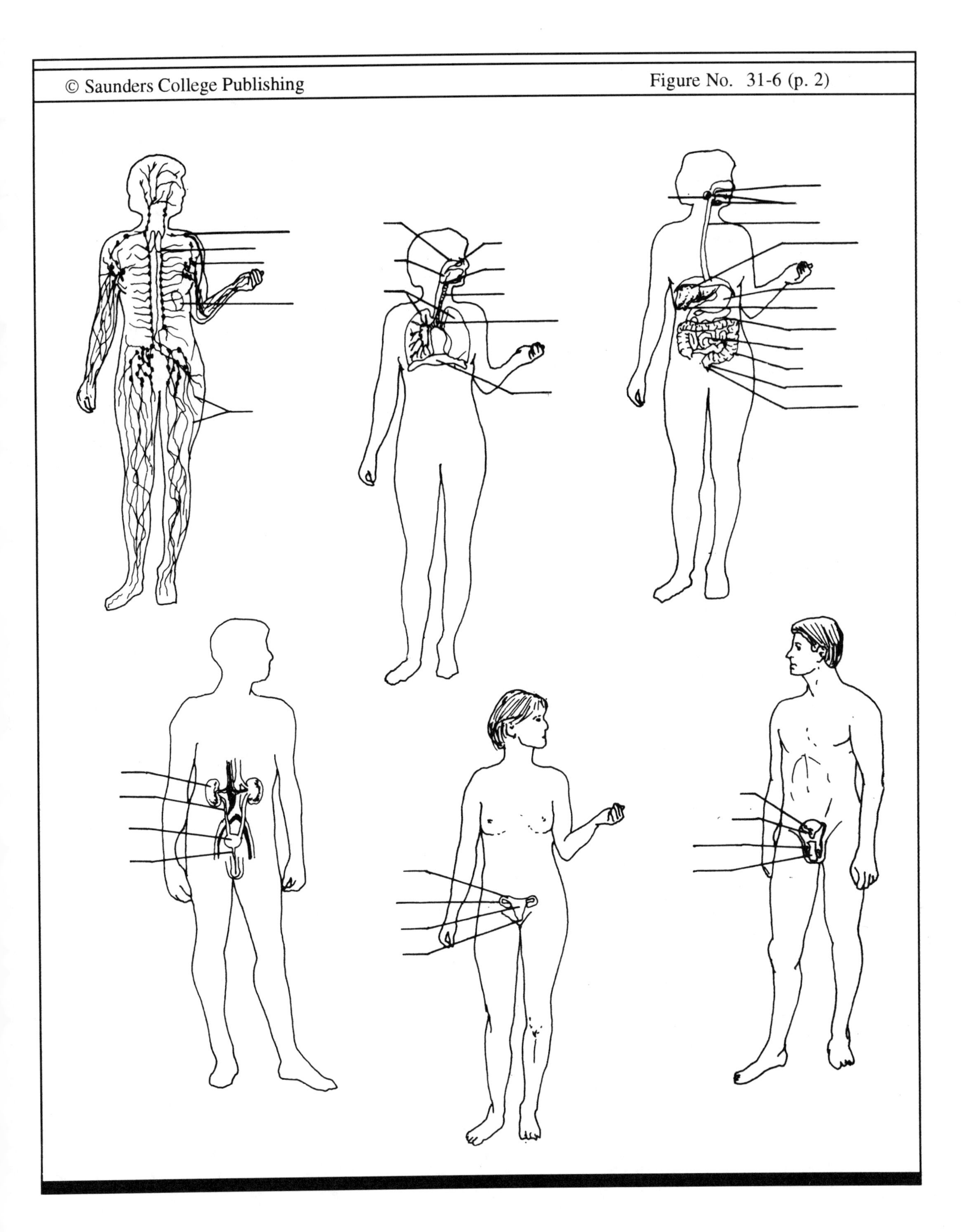

 Figure No. 32-2

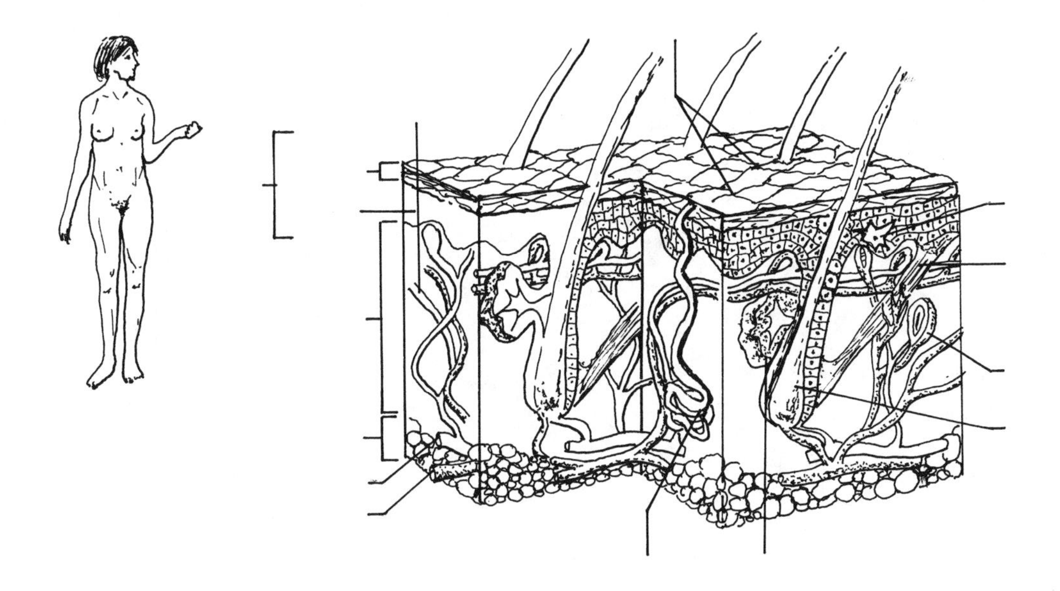

Figure No. 32-6

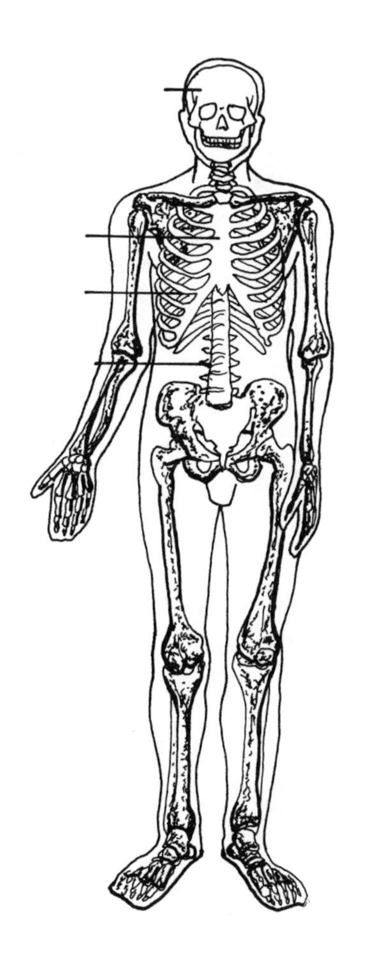

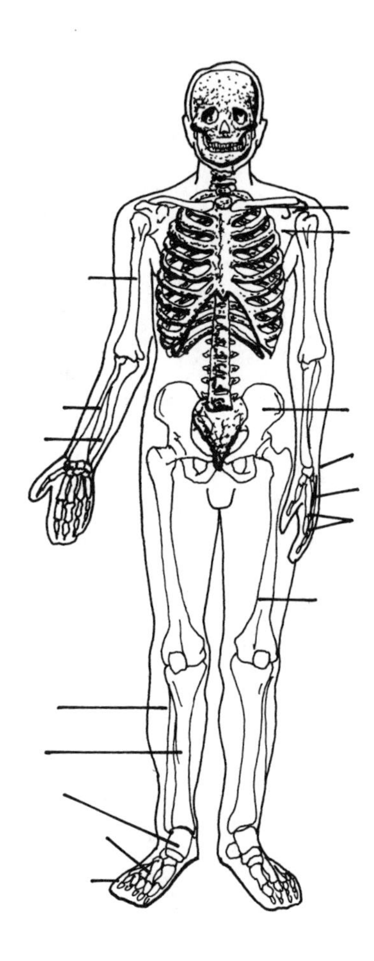

 Figure No. 32-8

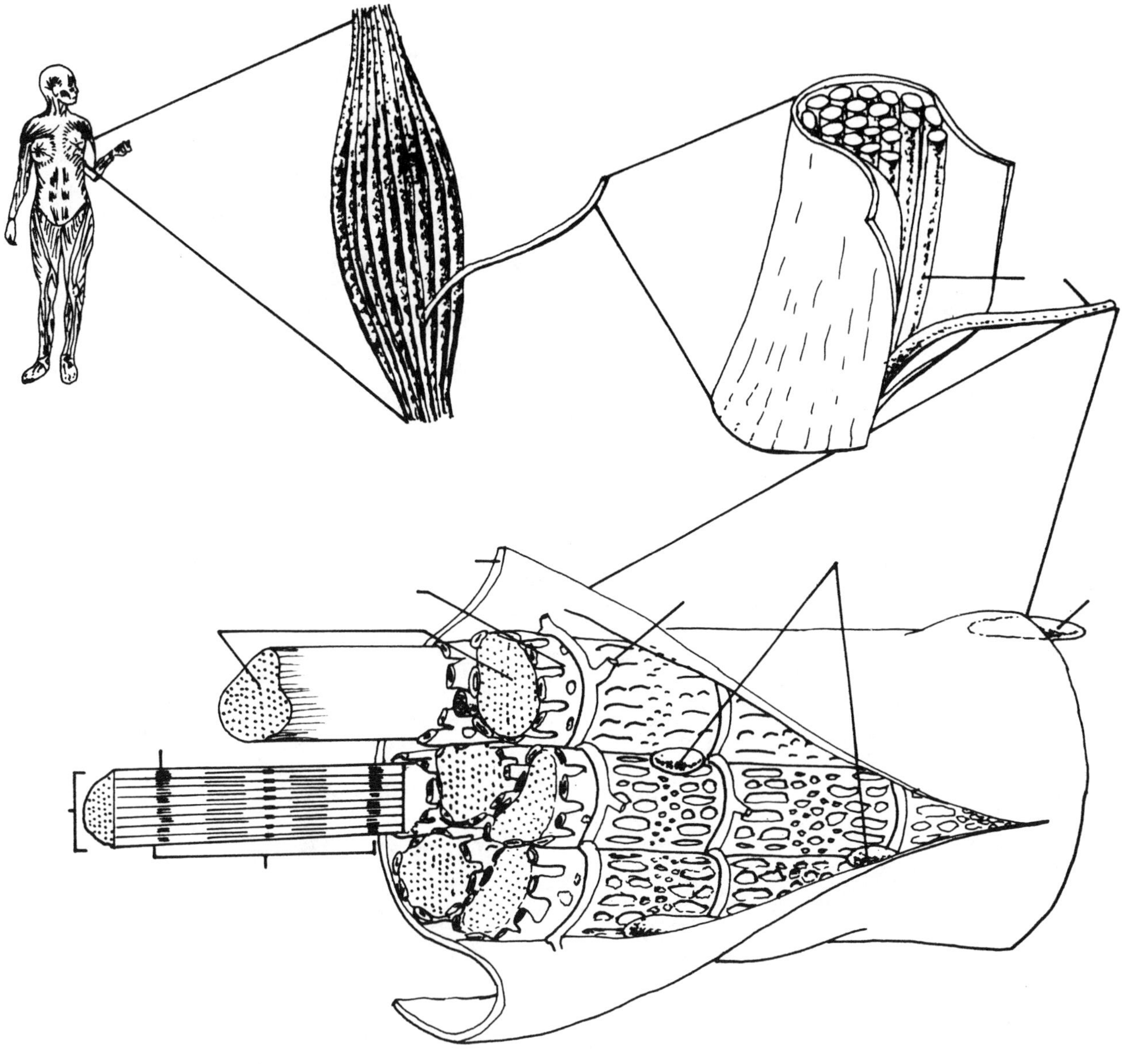

 Figure No. 32-10

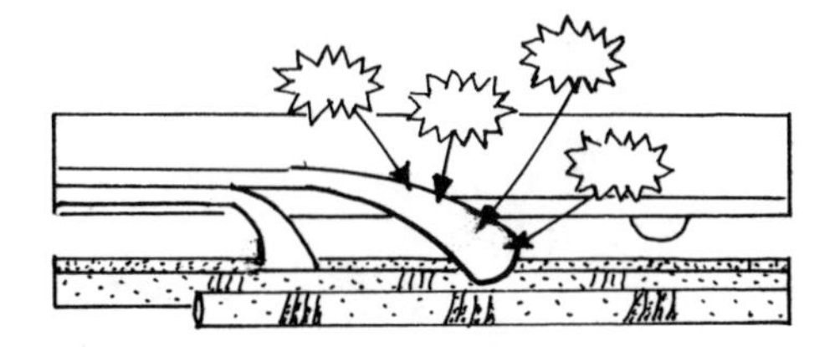

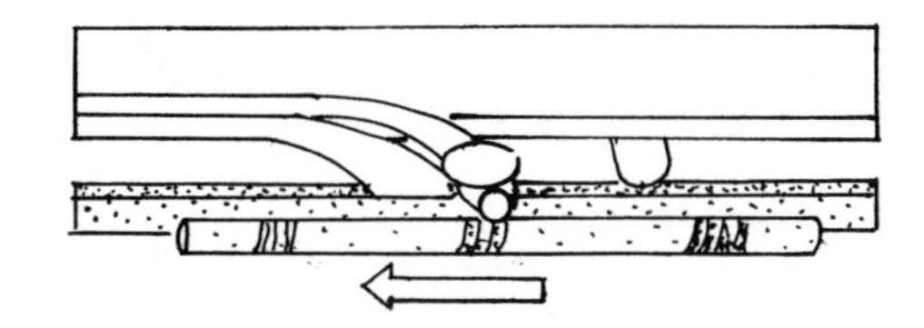

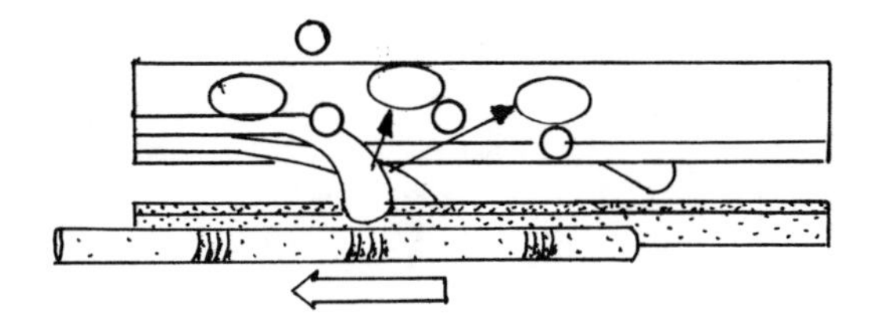

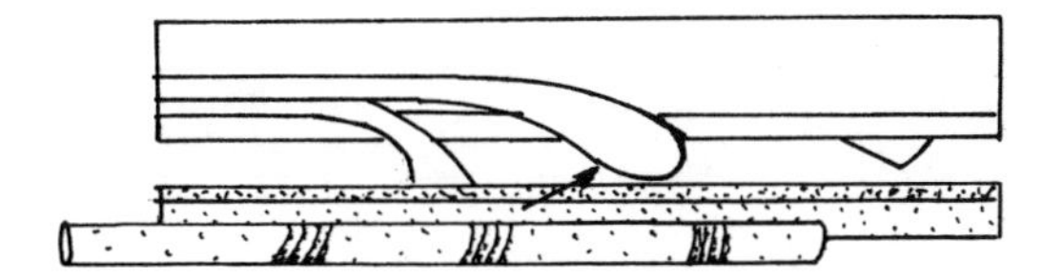

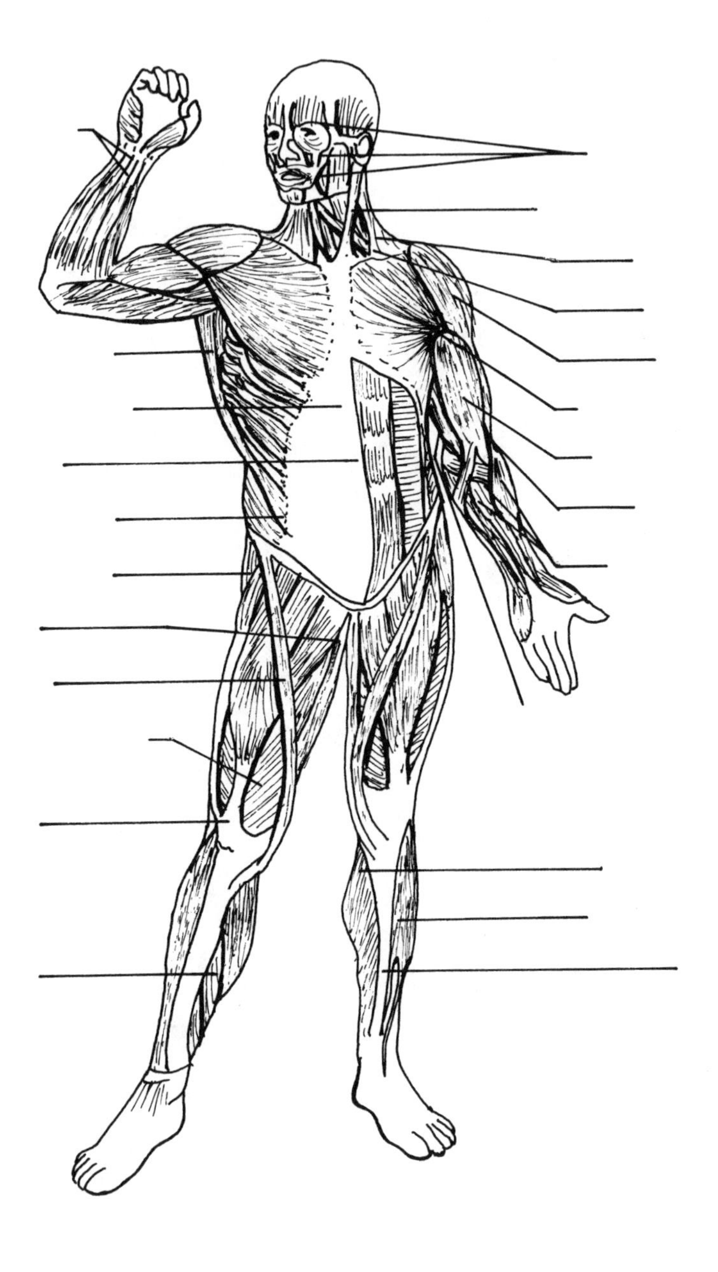

Figure No. 32-14b

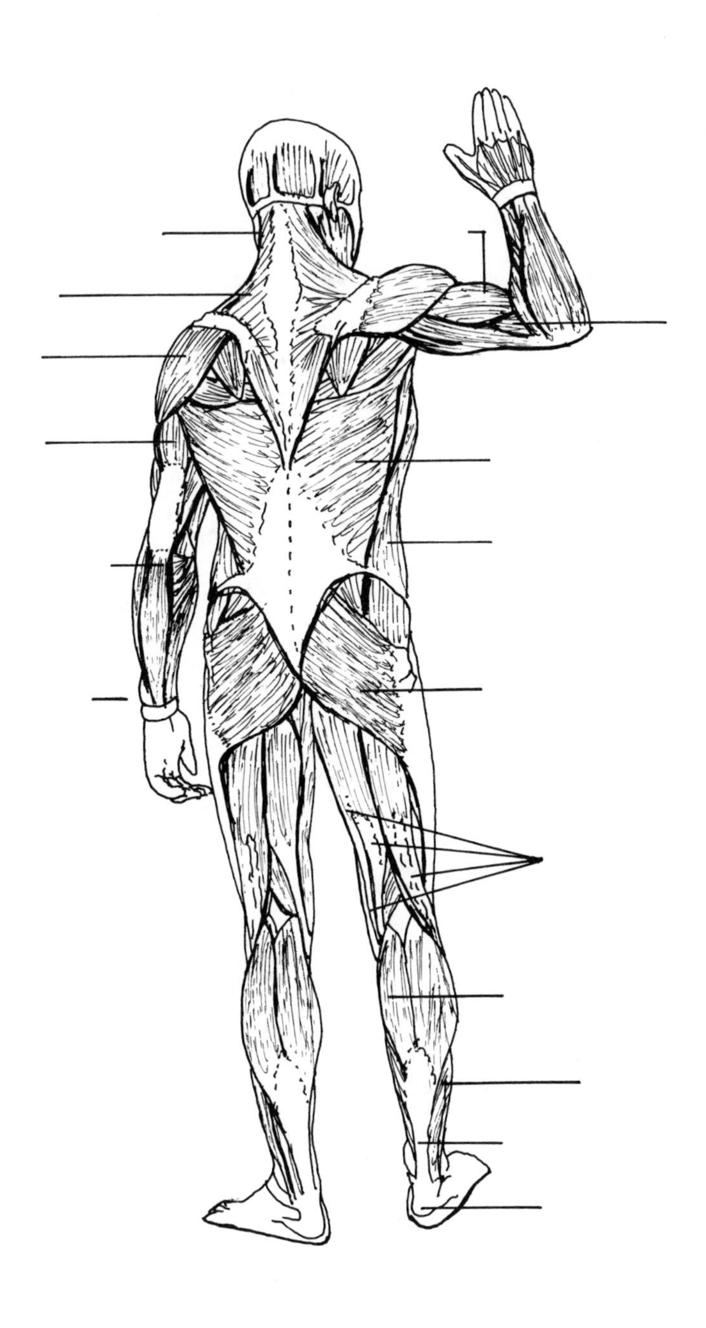

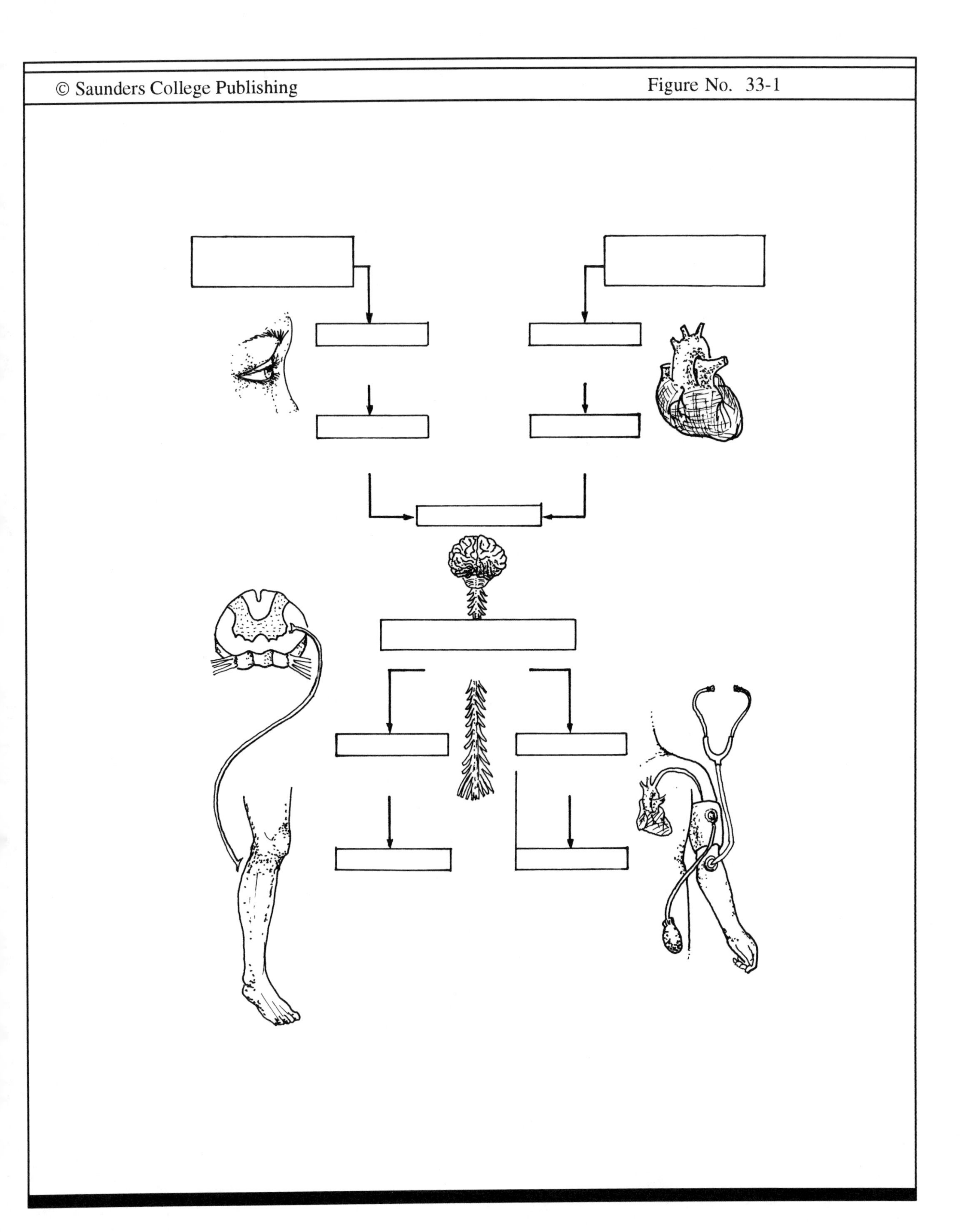
© Saunders College Publishing
Figure No. 33-1

Figure No. 33-5

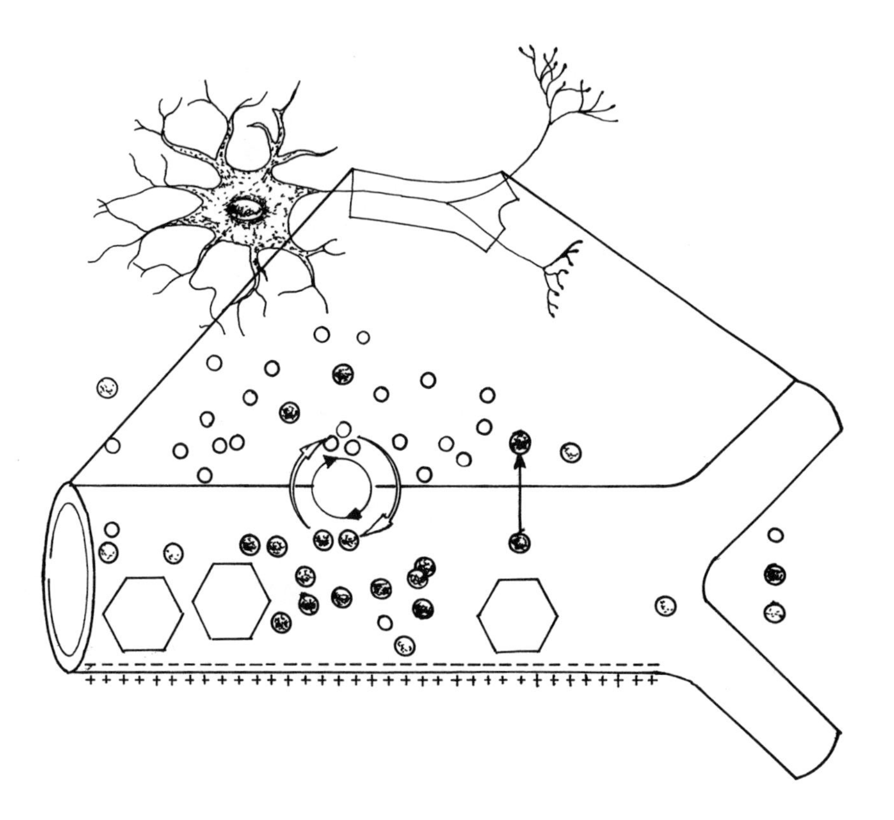

 Figure No. 33-9

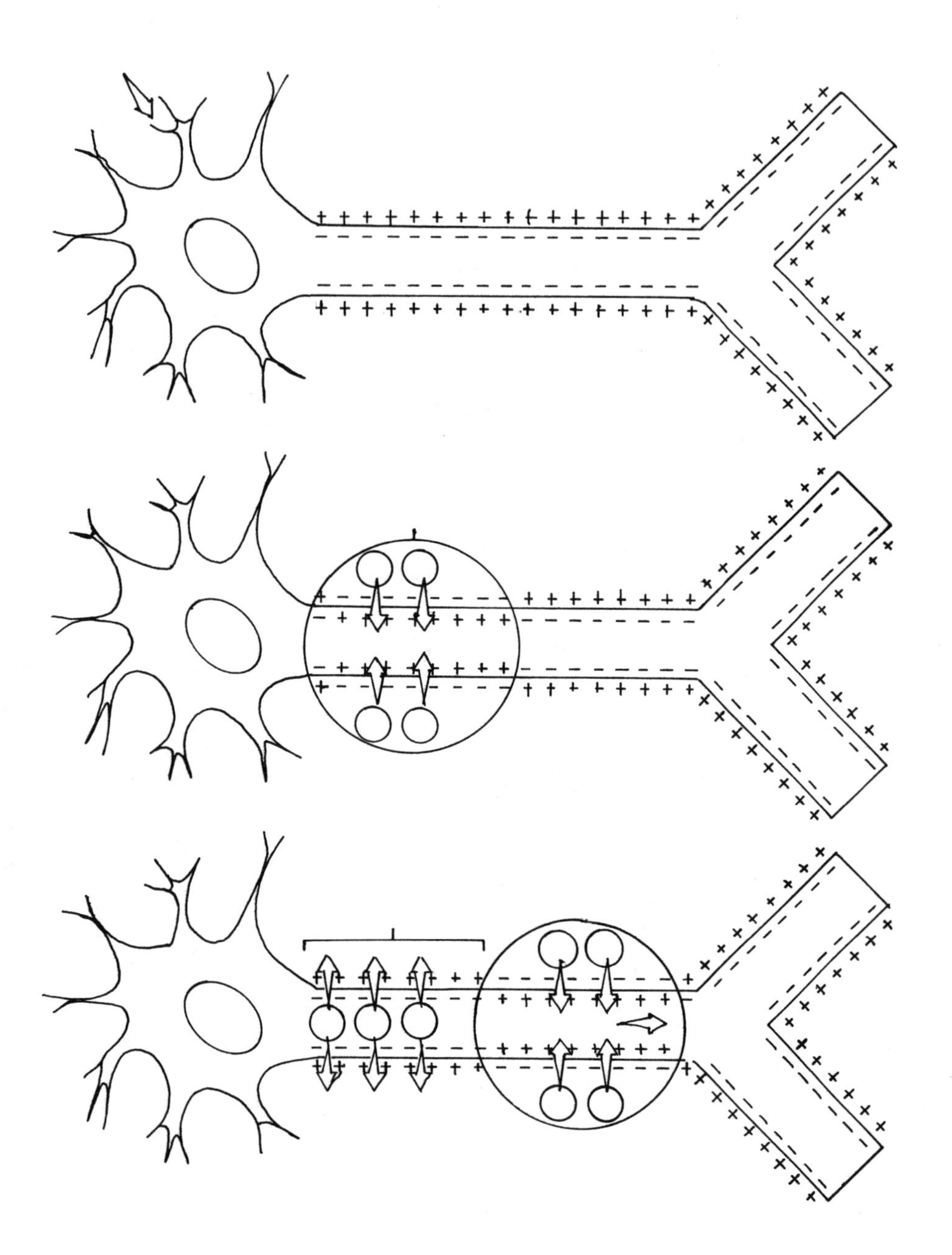

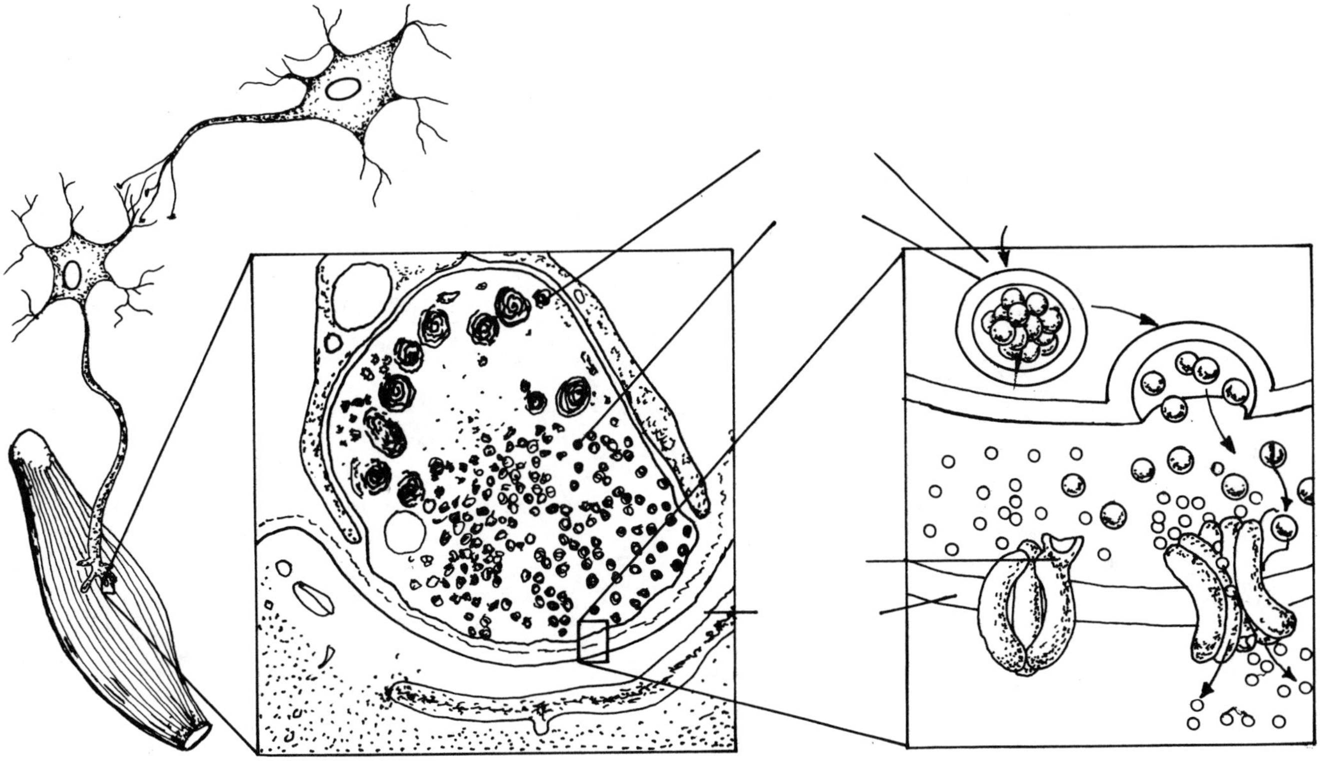

Figure No. 34-6

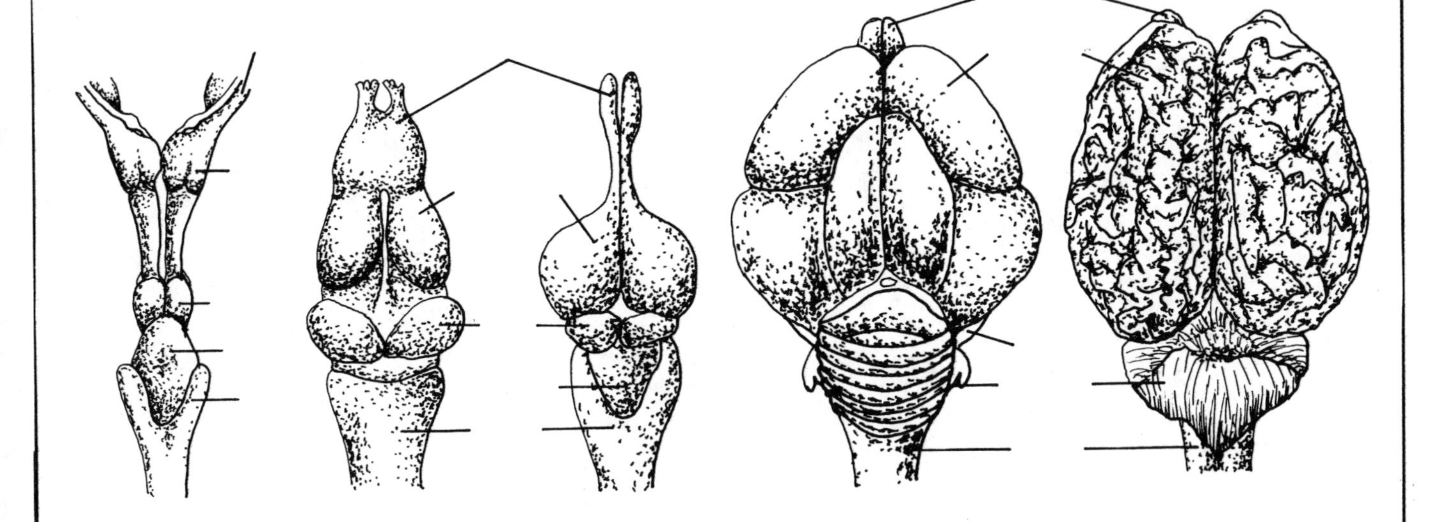

Figure No. 34-7

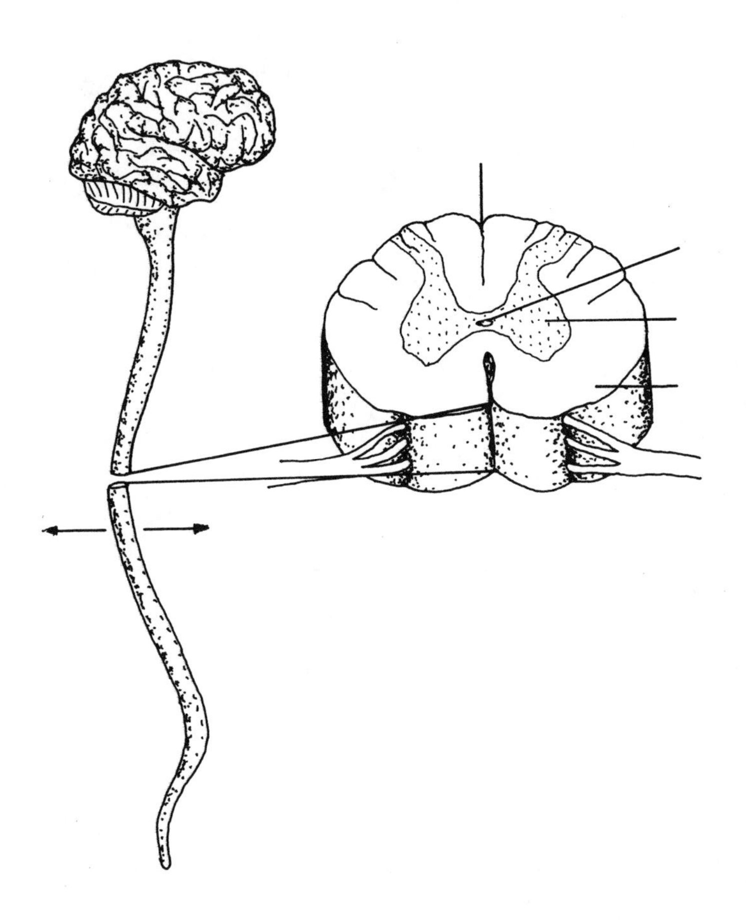

 Figure No. 34-8

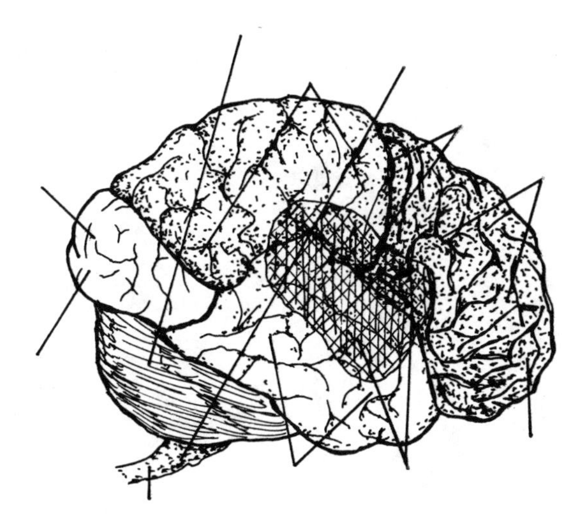

 Figure No. 34-9

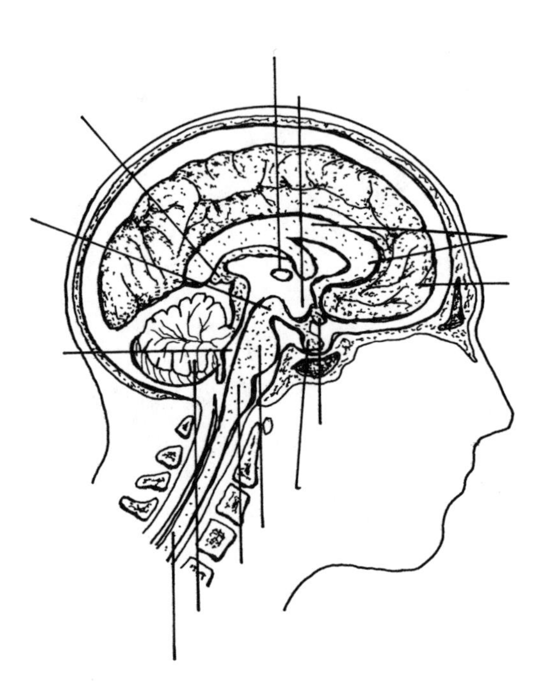

Figure No. 35-15

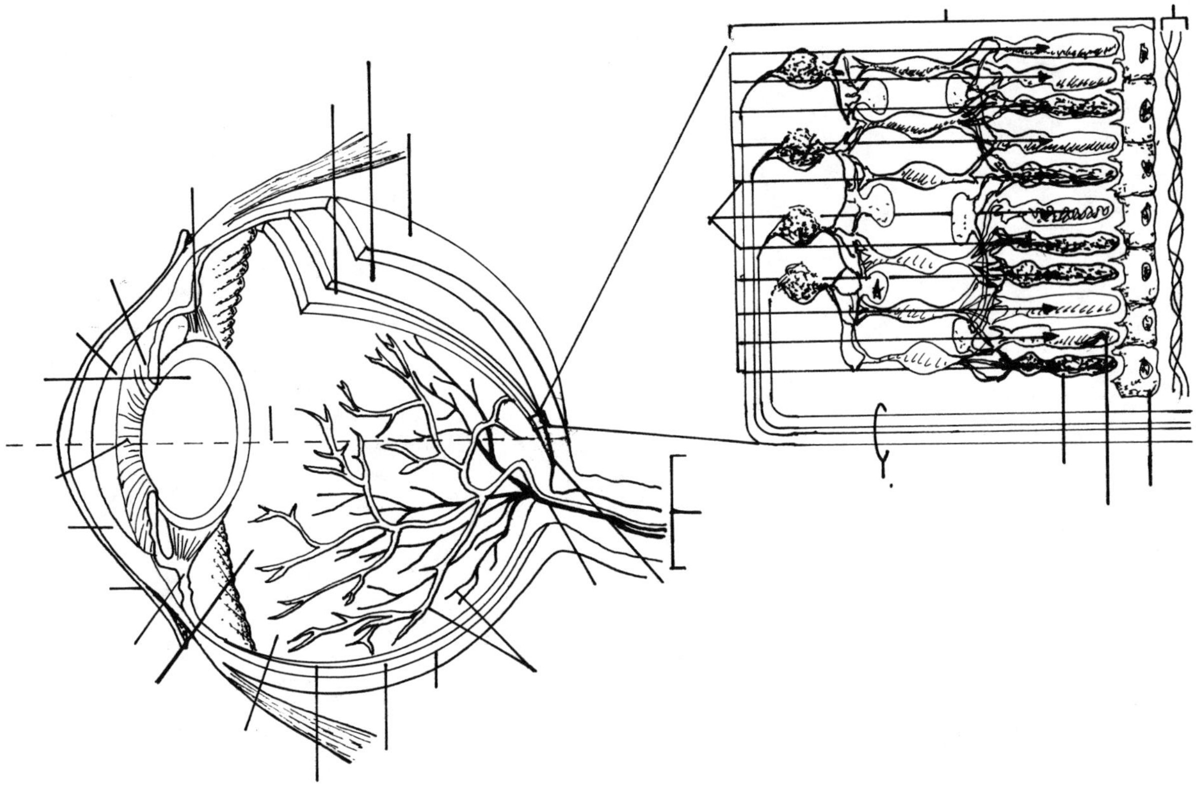

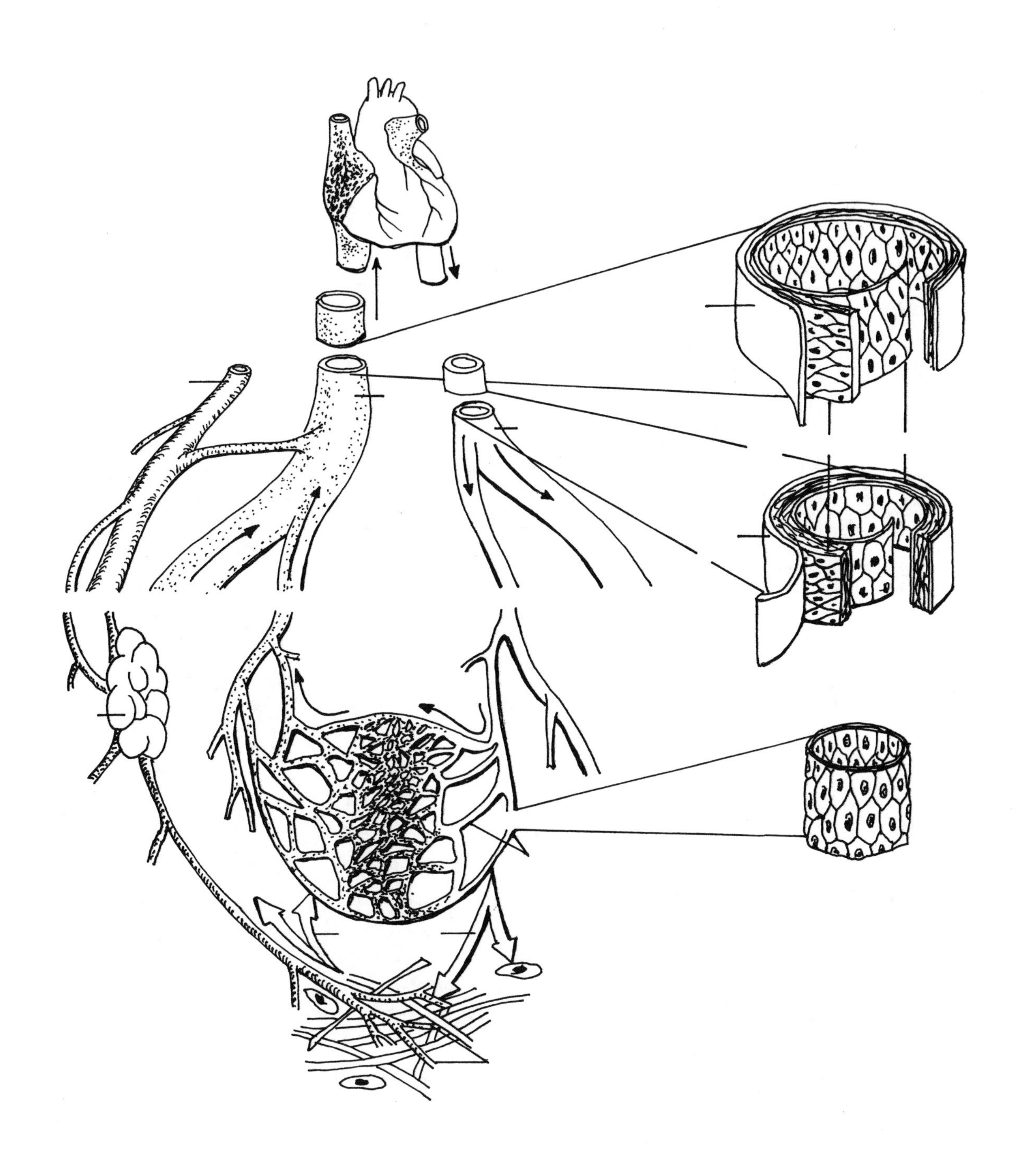

Figure No. 36-9

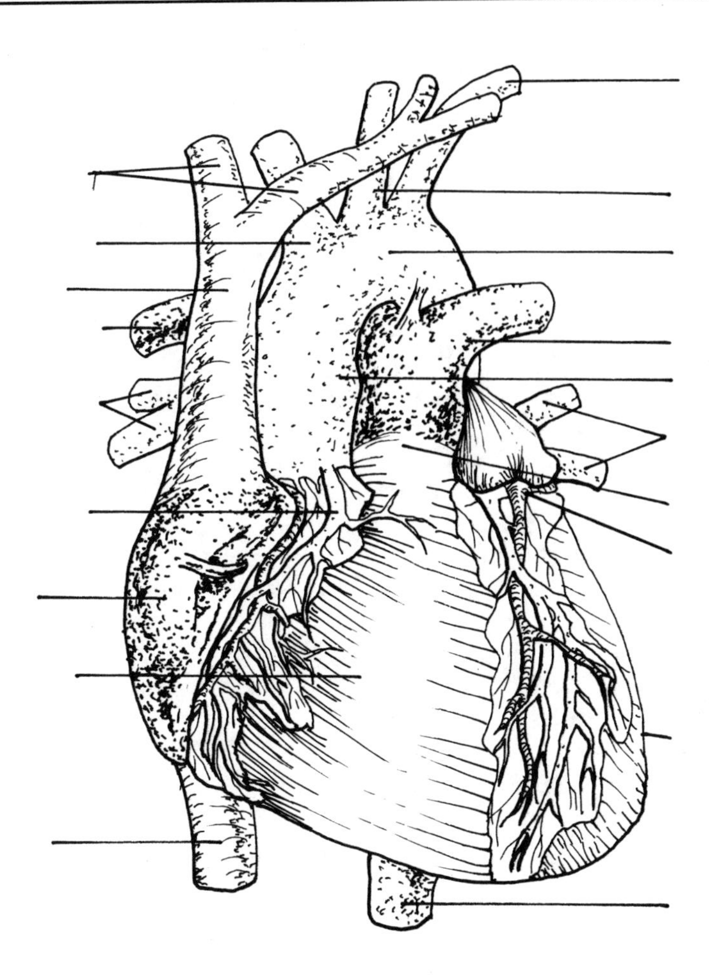

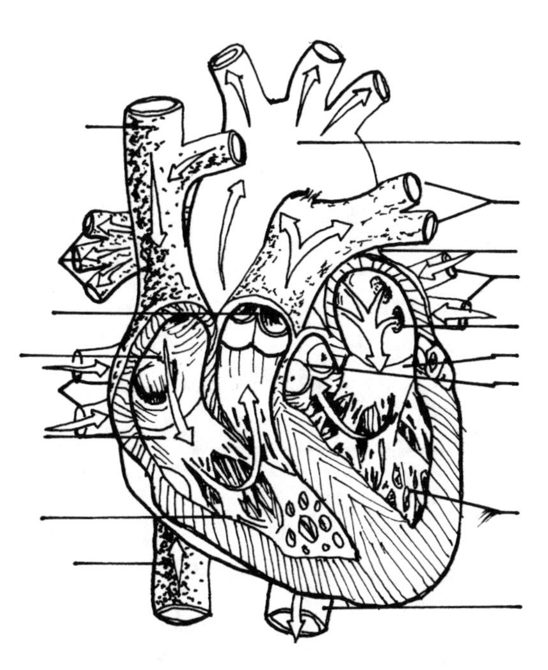

 Figure No. 36-15

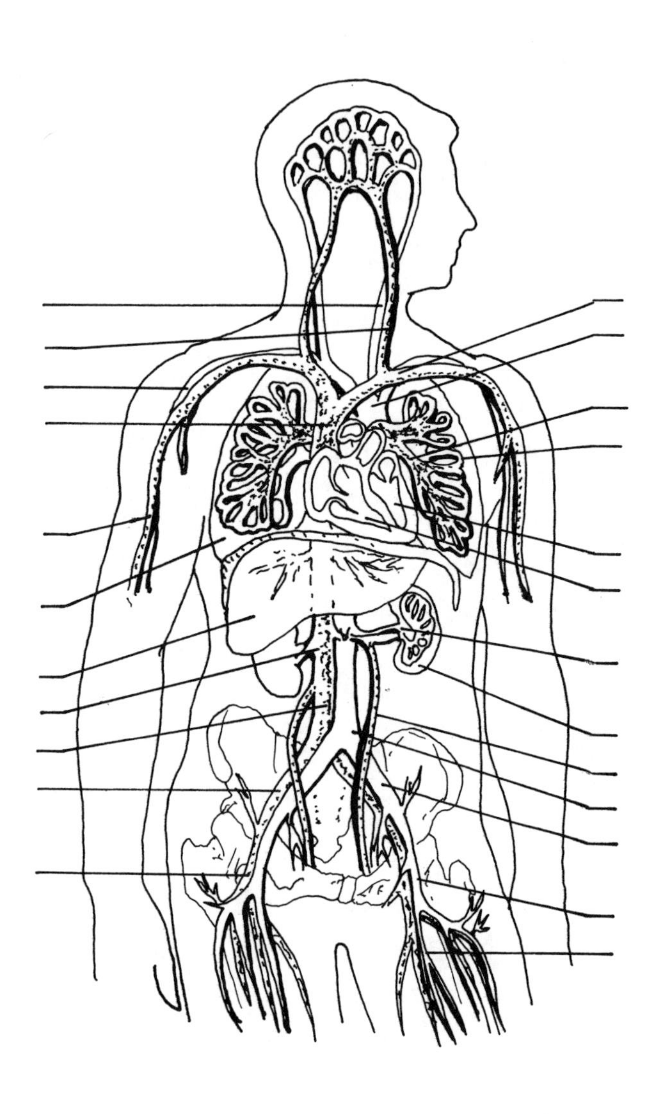

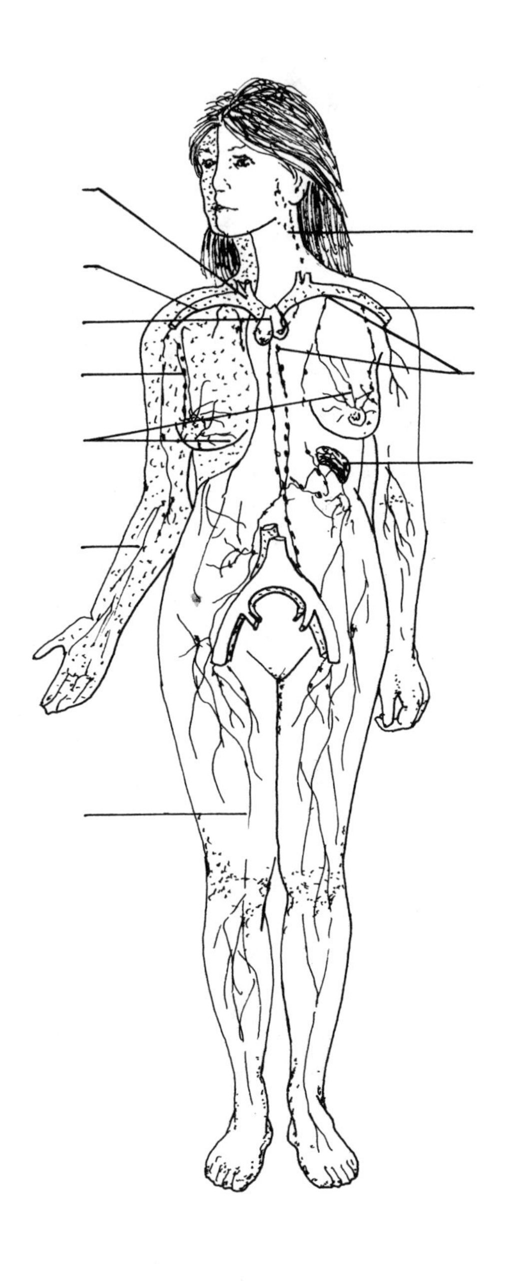

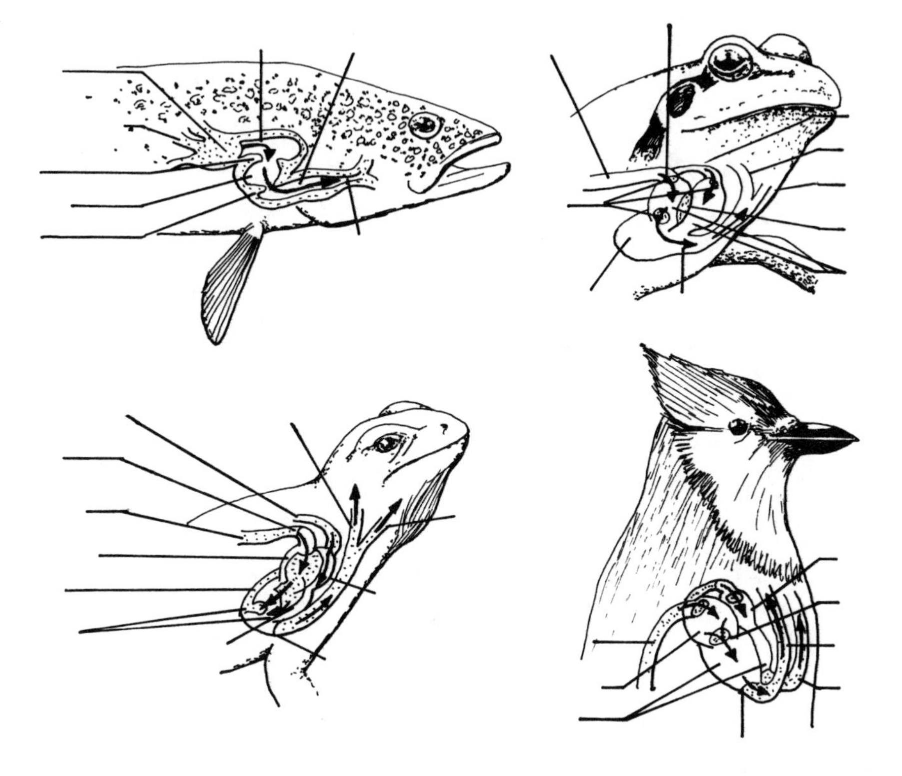

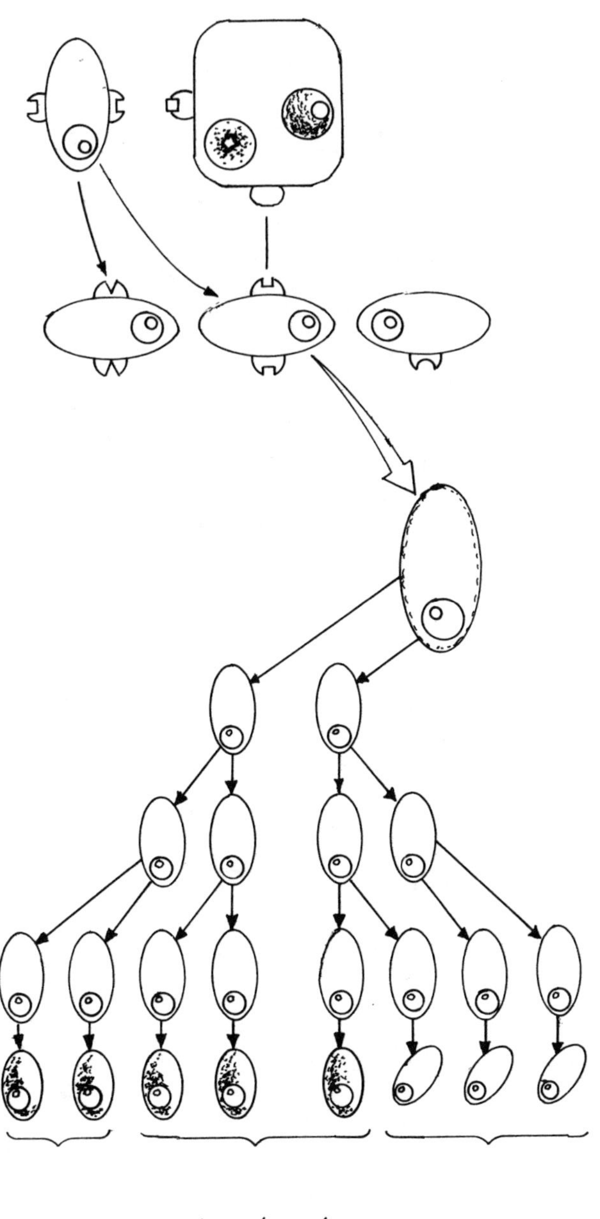

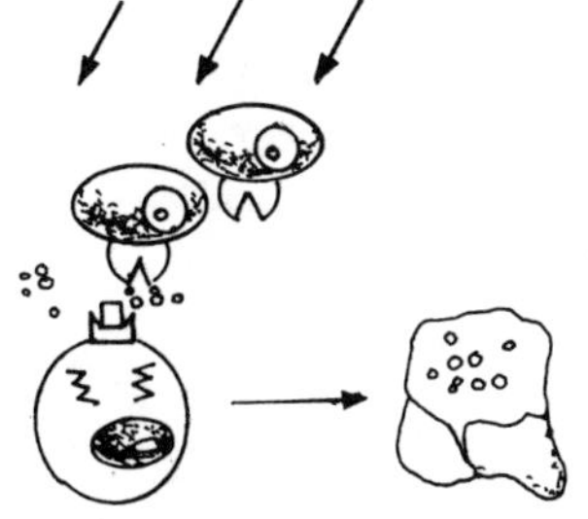

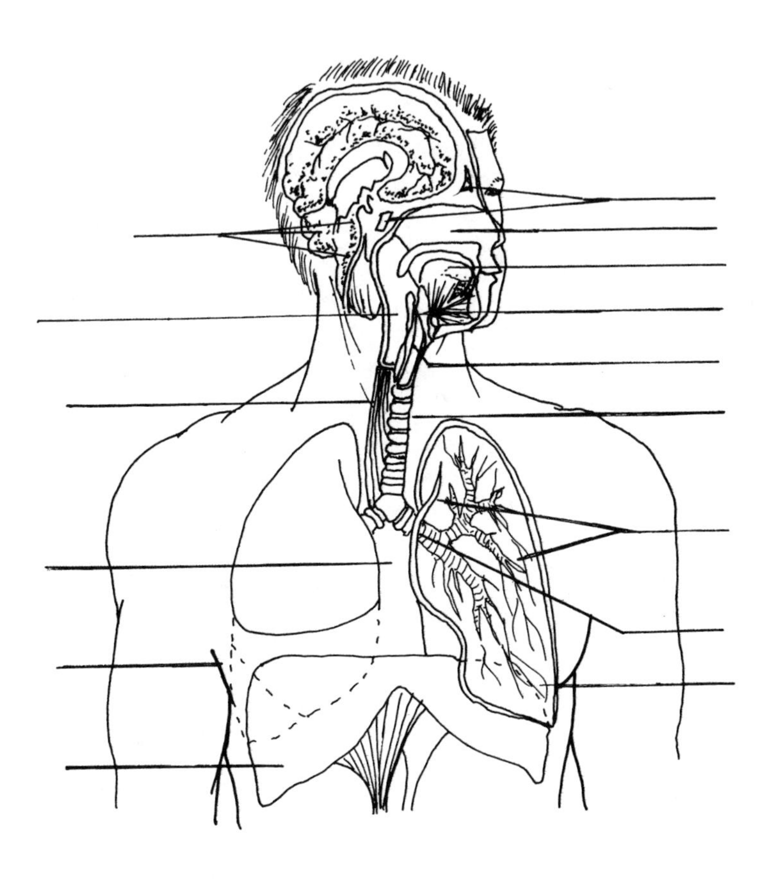

 Figure No. 38-7

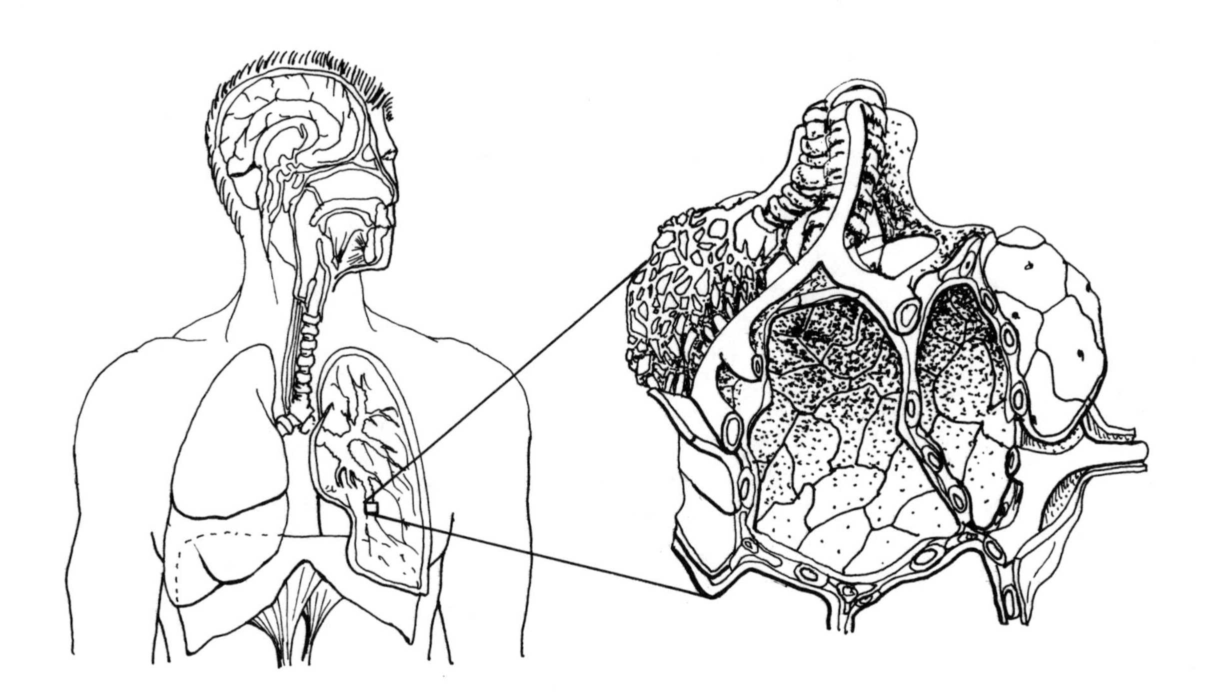

 Figure No. 39-4

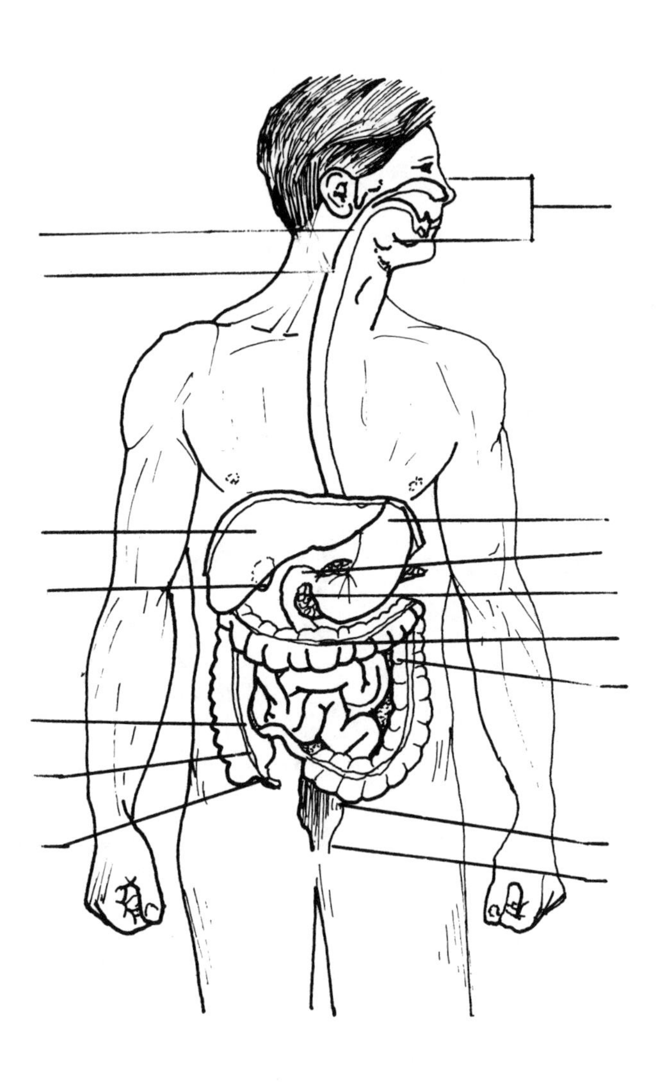

Figure No. 39-5

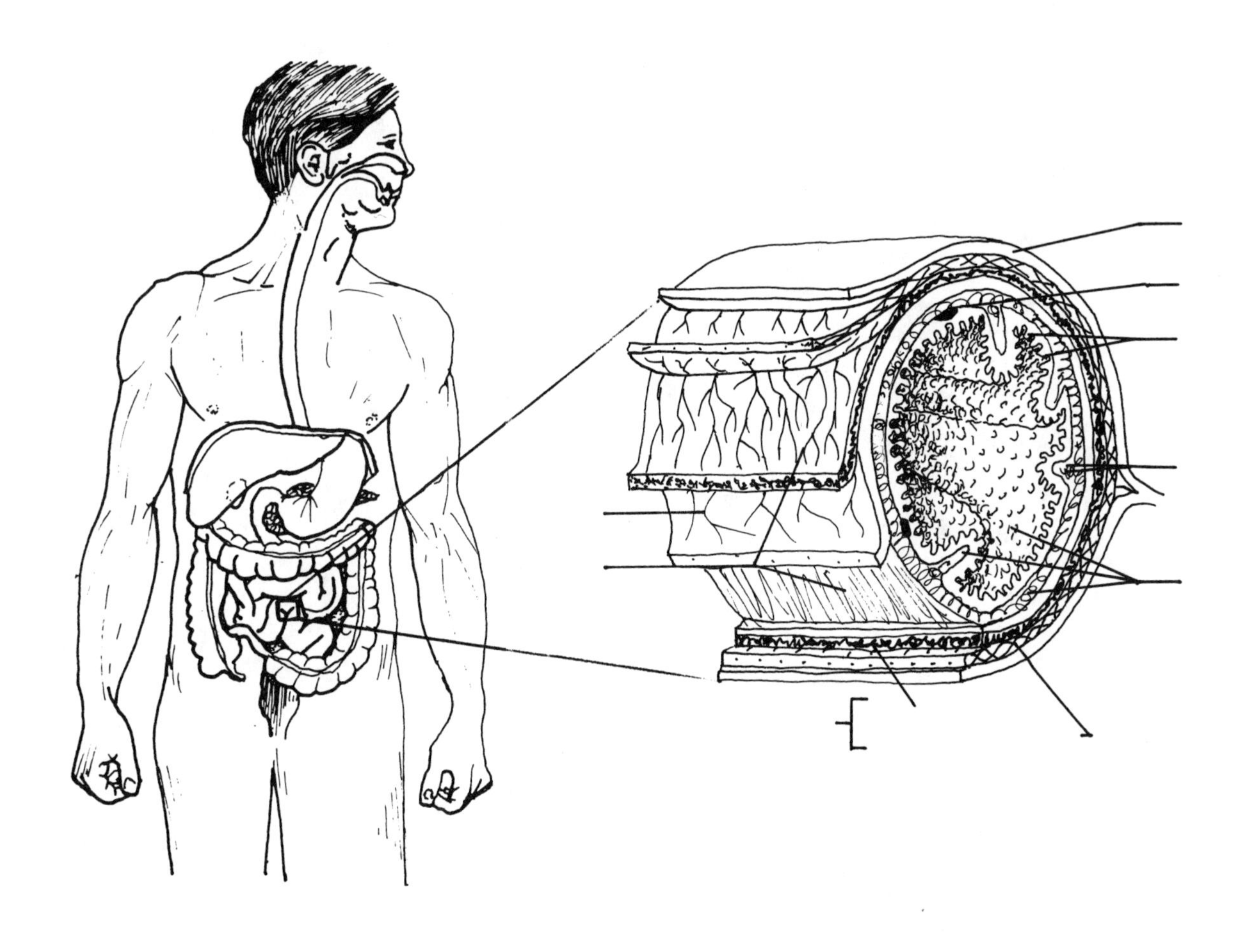

Figure No. 39-7

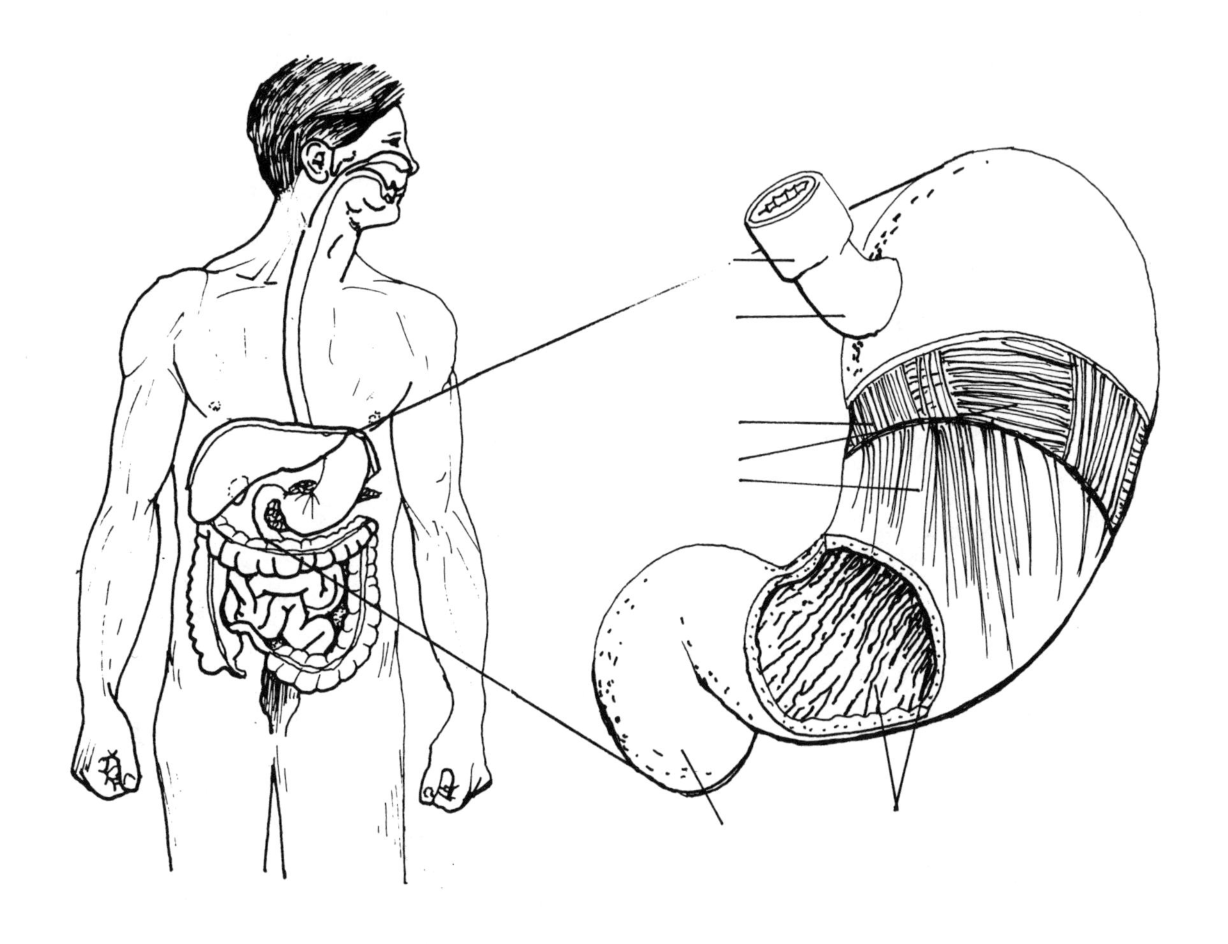

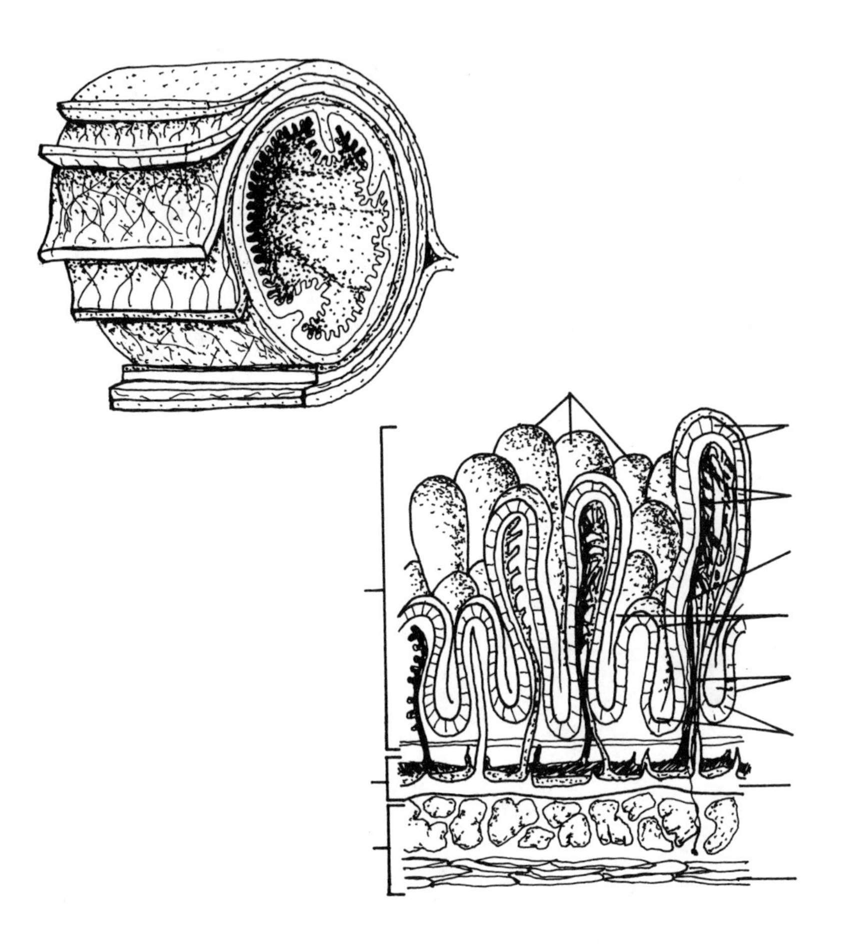

Figure No. 40-2

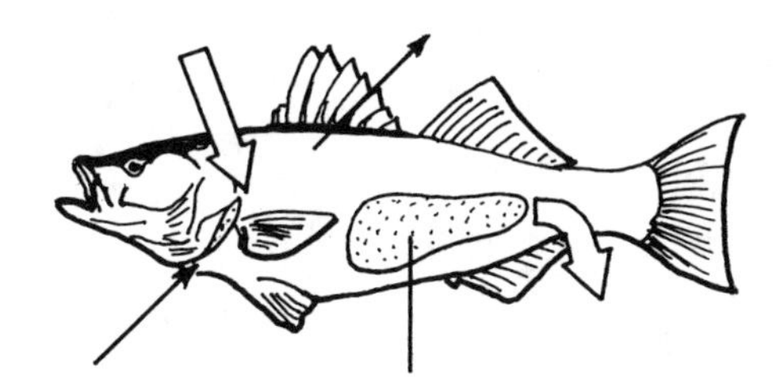

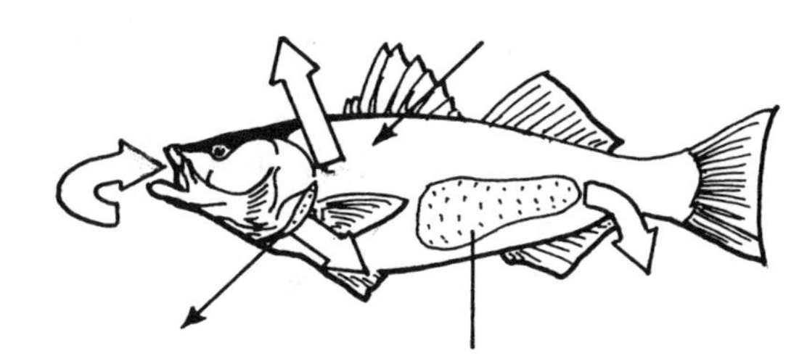

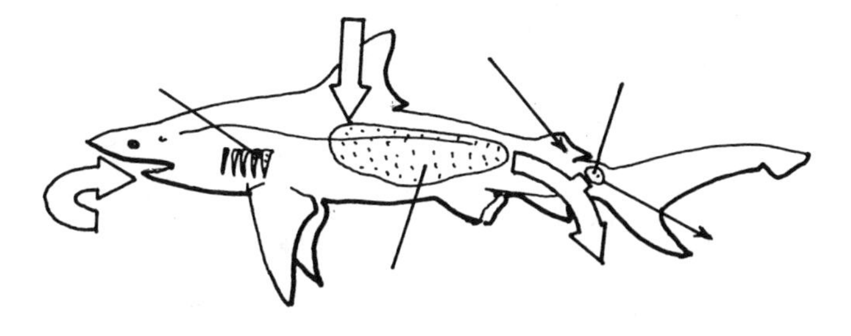

Figure No. 40-5

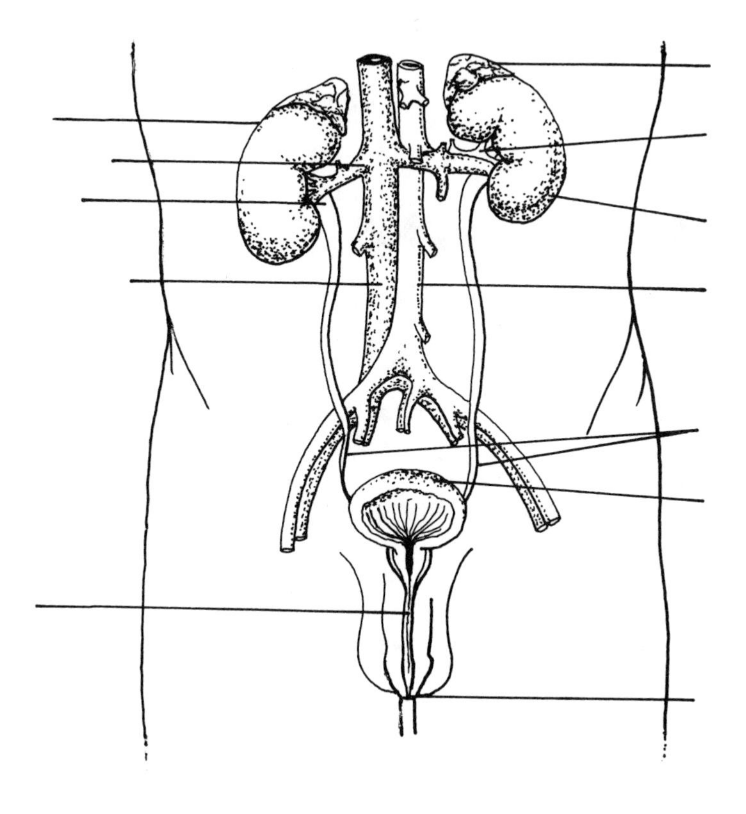

Figure No. 40-6

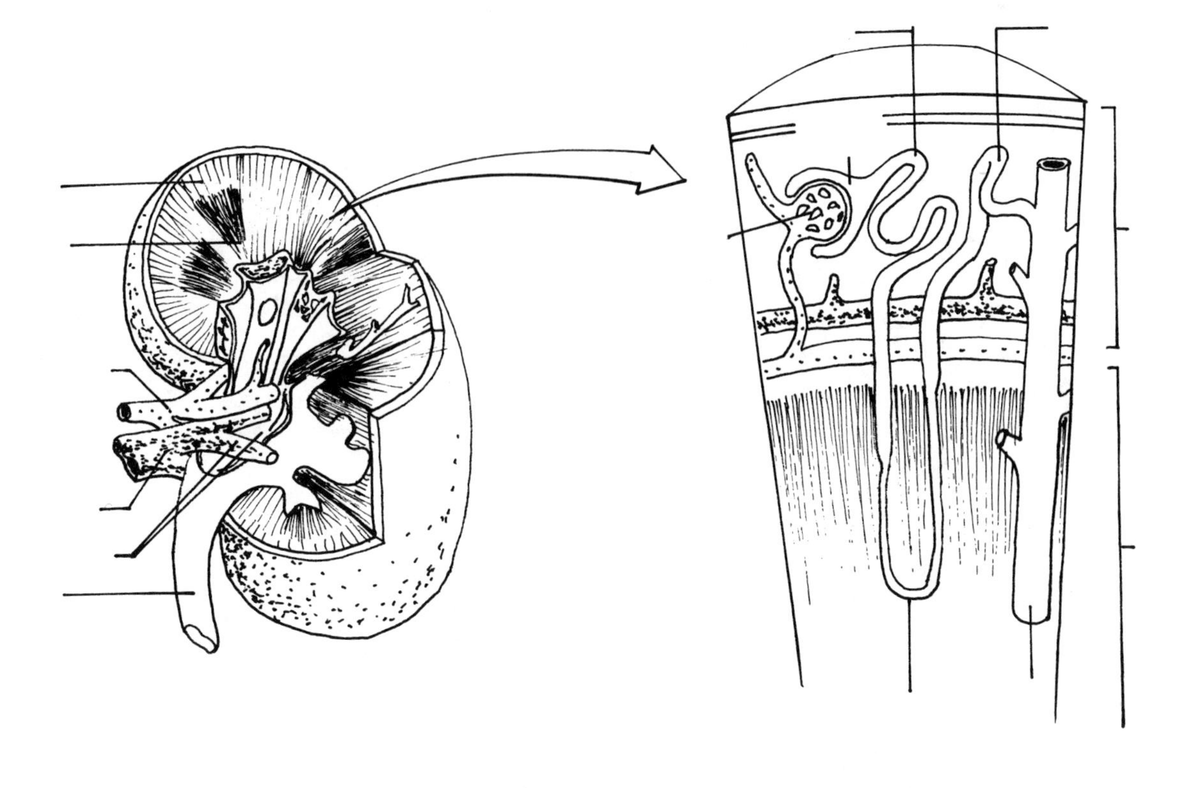

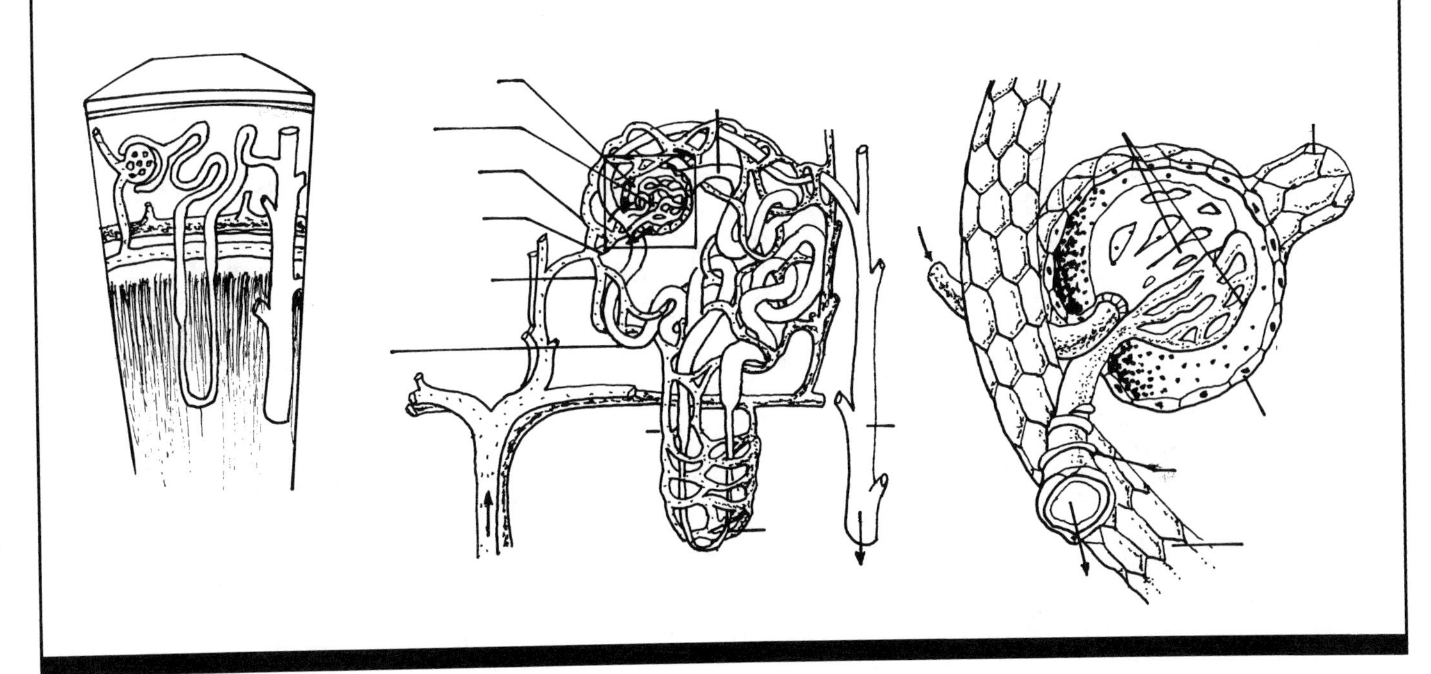

Figure No. 41-4

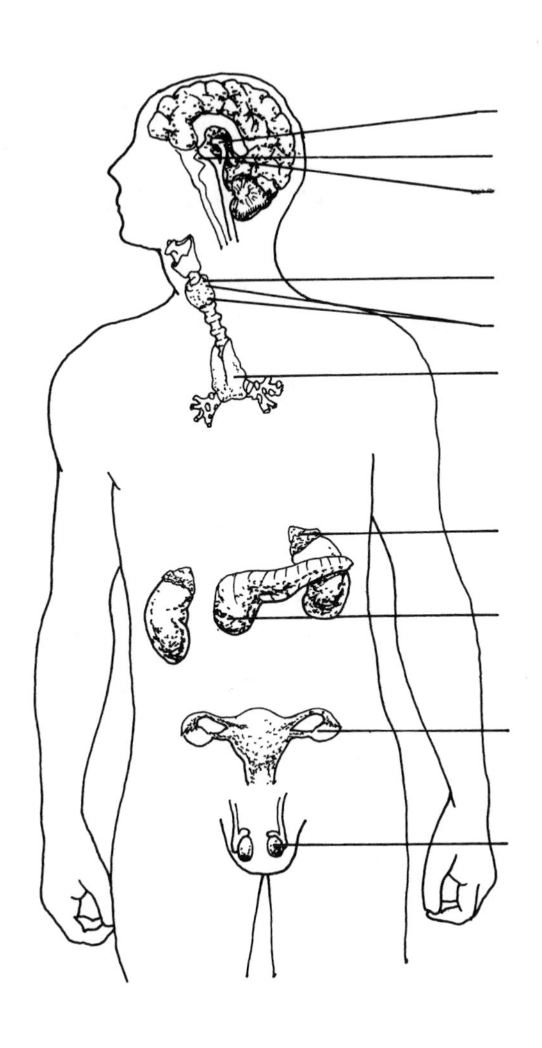

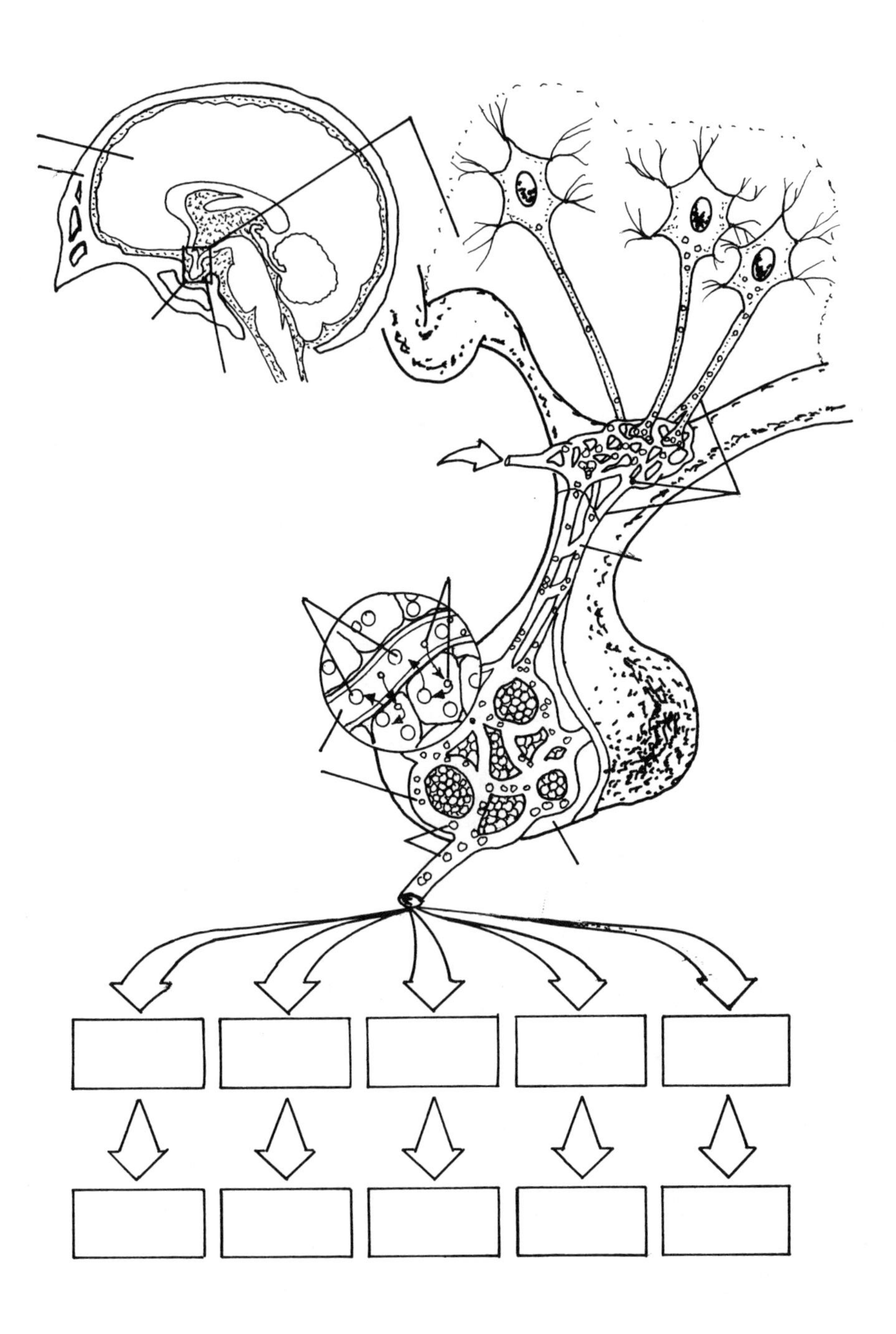

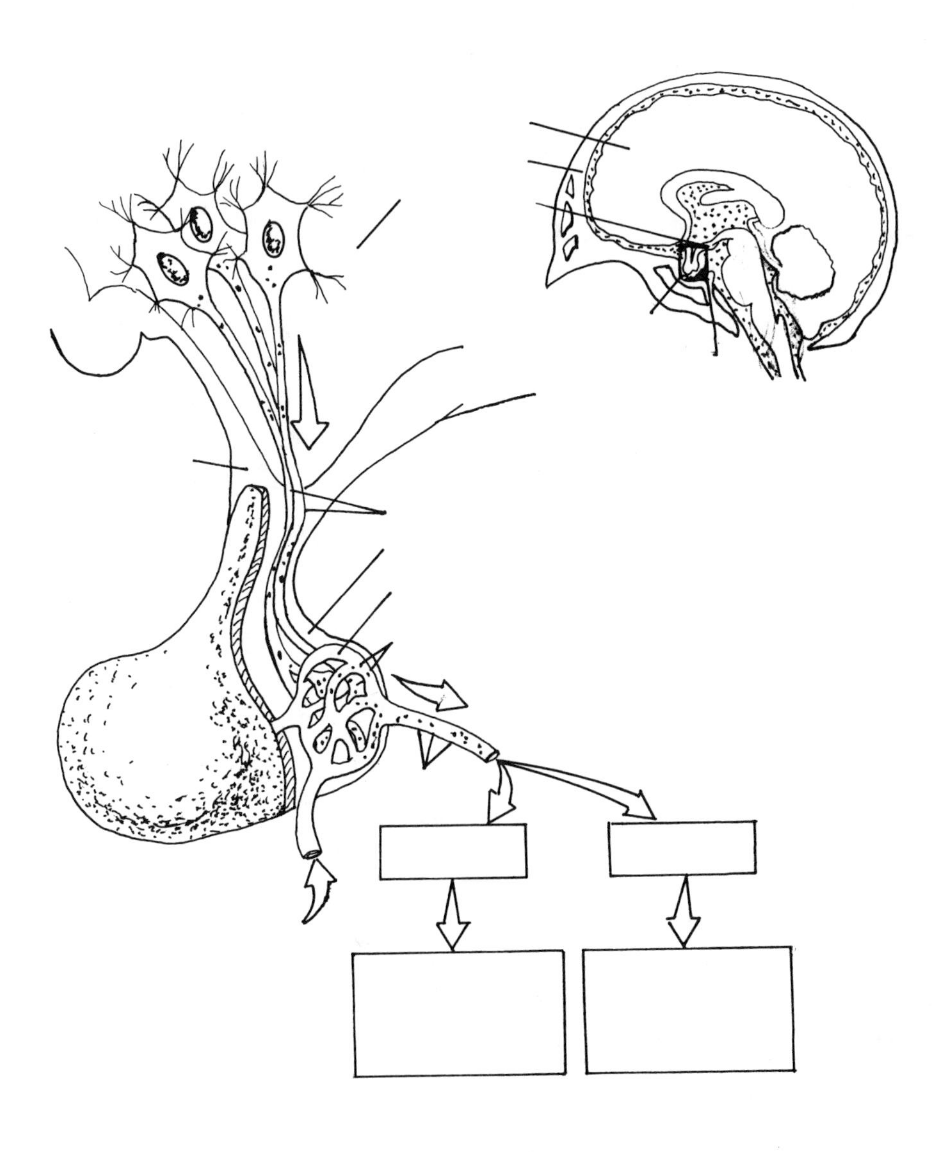

Figure No. 42-3

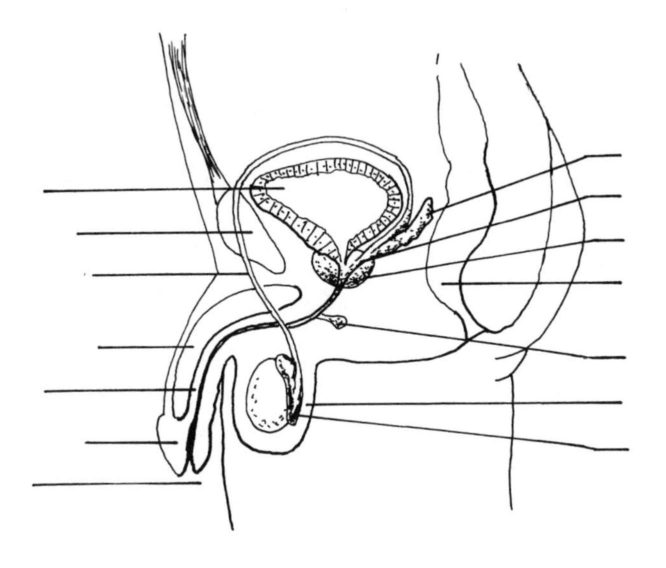

Figure No. 42-10

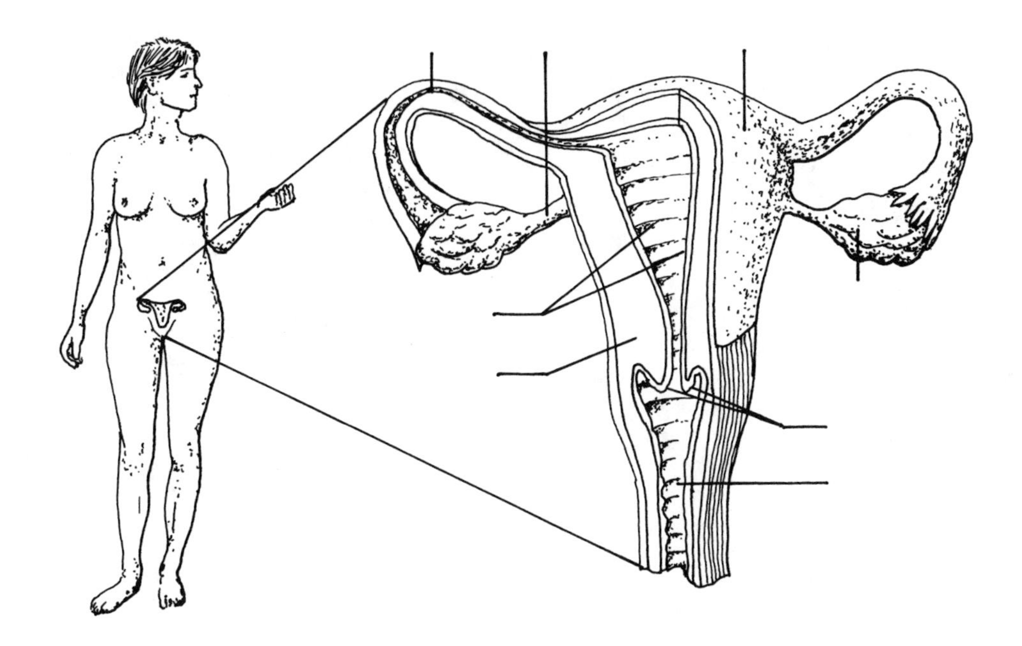

Figure No. 43-11

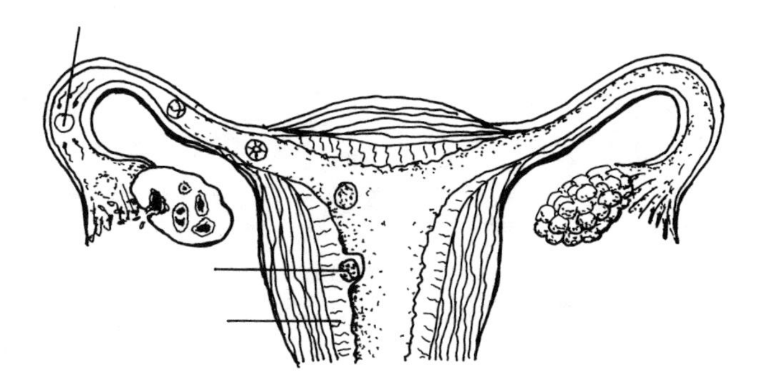

Figure No. 47-6

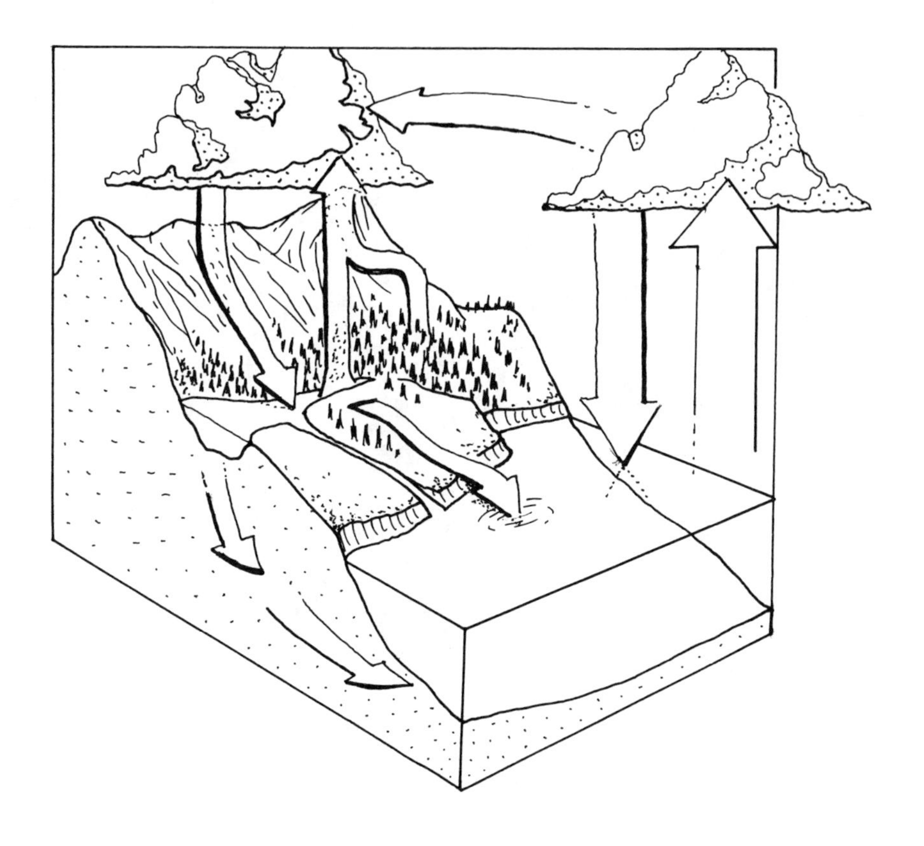